Bringing Forth Prosperity:
Capacity Innovation in Africa

RIVER PUBLISHERS SERIES IN MULTI BUSINESS MODEL INNOVATION, TECHNOLOGIES AND SUSTAINABLE BUSINESS

Series Editors

PETER LINDGREN
Aarhus University
Denmark

ANNABETH AAGAARD
Aarhus University
Denmark

Indexing: All books published in this series are submitted to Thomson Reuters Book Citation Index (BkCI), CrossRef and to Google Scholar.

The River Publishers Series in Multi Business Model Innovation, Technologies and Sustainable Business includes the theory and use of multi business model innovation, technologies and sustainability involving typologies, ontologies, innovation methods and tools for multi business models, and sustainable business and sustainable innovation. The series cover cross technology business modeling, cross functional business models, network based business modeling, Green Business Models, Social Business Models, Global Business Models, Multi Business Model Innovation, interdisciplinary business model innovation. Strategic Business Model Innovation, Business Model Innovation Leadership and Management, technologies and software for supporting multi business modeling, Multi business modeling and strategic multi business modeling in different physical, digital and virtual worlds and sensing business models. Furthermore the series includes sustainable business models, sustainable & social innovation, CSR & sustainability in businesses and social entrepreneurship.

Key topics of the book series include:

- Multi business models
- Network based business models
- Open and closed business models
- Multi Business Model eco systems
- Global Business Models
- Multi Business model Innovation Leadership and Management
- Multi Business Model innovation models, methods and tools
- Sensing Multi Business Models
- Sustainable business models
- Sustainability & CSR in businesses
- Sustainable & social innovation
- Social entrepreneurship and -intrapreneurship

For a list of other books in this series, visit www.riverpublishers.com

Bringing Forth Prosperity: Capacity Innovation in Africa

Benjamin F. Bobo

Loyola Marymount University
Los Angeles, CA, USA

LONDON AND NEW YORK

Published 2017 by River Publishers
River Publishers
Alsbjergvej 10, 9260 Gistrup, Denmark
www.riverpublishers.com

Distributed exclusively by Routledge
4 Park Square, Milton Park, Abingdon, Oxon OX14 4RN
605 Third Avenue, New York, NY 10017, USA

Bringing Forth Prosperity: Capacity Innovation in Africa / by Benjamin F. Bobo.

© 2017 River Publishers. All rights reserved. No part of this publication may be reproduced, stored in a retrieval systems, or transmitted in any form or by any means, mechanical, photocopying, recording or otherwise, without prior written permission of the publishers.

Routledge is an imprint of the Taylor & Francis Group, an informa business

ISBN 978-87-93519-29-9 (print)

While every effort is made to provide dependable information, the publisher, authors, and editors cannot be held responsible for any errors or omissions.

To Patricia, Benjelani and Yohancé

Contents

Preface   xi

Acknowledgements   xvii

List of Exhibits   xix

1 Introduction   1
   1.1 Direction and Purpose of Book 2
   1.2 Need and Approach 4
   1.3 Objectives 5
   1.4 Undertaking the Objectives 6

2 Capital Theory and Development: Has Africa Been Overlooked?   9
   2.1 Capital Theory and Government Corruption 11
      2.1.1 Government Institutions 14
      2.1.2 Give-and-Take 16
      2.1.3 The Government Apparatus: Governmental Decree 18
      2.1.4 Constitution 18
      2.1.5 Rule of Law 20
      2.1.6 Bodies of Government 20
      2.1.7 So – Has Africa Been Overlooked? 20

3 Towards an Analytical Policy Framework for Assessing Capacity Innovation: A Standalone African Perspective   23
   3.1 Africa's Predicament 25
   3.2 Capacity Innovation's Purpose: What It Needs to Accomplish 27
      3.2.1 What's Missing – A Look at the Core Challenges 36

	3.2.2	Comparative Advantage	37
	3.2.3	Mini-States and Critical Mass	38
	3.2.4	Price-Takers	41
	3.2.5	Economic Complementarity	42
	3.2.6	Communication	43
	3.2.7	Collective Self-Reliance – Economic Integration	44
	3.2.8	Currency	45
	3.2.9	Governance	47
	3.2.10	The Capacity Innovation Endgame	49
	3.2.11	What Might Be Done: The Path to Capacity Innovation	52
	3.2.12	Closing Comments	55

4 Top Down–Bottom Up Capacity Innovation 57
4.1 Integrative Approach: Top Down–Bottom Up Convergence 69
4.2 Capacity Innovation for Prosperity: Top Down–Bottom Up with a Twist 71
4.3 Land Tenure, Property Rights and Ownership 75

5 The Country Capacity ID 83
5.1 Population 98
5.2 Resources of Commercial Importance 98
5.3 Literacy 99
5.4 Human Development Index 101
5.5 Capital Formation 102
5.6 Infrastructure 105
5.7 Institutions – Rule of Law 107
5.8 Citizen Participation 108
5.9 Foreign Direct Investment (FDI) 109
5.10 Military 111
5.11 Closing Observations 115

6 Africa-MNC Strategic Alliance 117
6.1 Opportunistic Behavior 120
6.2 Opportunistic Balancing 121
6.3 Developing a Mindset 122
6.4 The Corruption Dilemma 124

	6.5	Corruption, Human Rights Violations and Crimes Against Humanity: Intersecting Acts	127
	6.6	Interdicting Corruption: The Trump Card	134

7 Path to Capacity Innovation: An Africa-MNC Strategic Alliance — 137

- 7.1 The Linking Process ... 141
  - 7.1.1 Institutional Component ... 141
  - 7.1.2 Market Component ... 142
  - 7.1.3 World Bank/IMF ... 146
  - 7.1.4 Non-Governmental Organizations ... 147
  - 7.1.5 Foreign Direct Investment ... 149
  - 7.1.6 Created Comparative Advantage ... 150
  - 7.1.7 Regional Integration ... 152
  - 7.1.8 Economic Complementarity ... 154
  - 7.1.9 Price-Maker Status ... 156
  - 7.1.10 Integrated Telecommunications ... 158
  - 7.1.11 Infrastructure Development ... 159
  - 7.1.12 NEPAD-MNC-NGO Convergence ... 163
  - 7.1.13 Critical Mass ... 164
  - 7.1.14 Transfer Pricing/Trade Misinvoicing ... 166
  - 7.1.15 Foreign Direct Investment Fund ... 172
  - 7.1.16 NEPAD Bank ... 175
  - 7.1.17 African Security Force-FDI Protection ... 176
  - 7.1.18 Country Capacity ID ... 177
  - 7.1.19 Organizational Tie-in ... 178
  - 7.1.20 Unlocking the Lock – Endgame ... 180

8 African Continental Security Apparatus: Accommodating FDI — 185

- 8.1 Special Fund ... 192

9 Epilogue — 195

Appendix — 199

Bibliography — 227

Index — 253

About the Author — 263

Preface

An invasion of armies can be resisted, but not an idea whose time has come.
—Victor Hugo (1802–1885)[1]

In the time since my first major exposition on Africa's economic development more than three decades past—a book draft[2] embedded in Third World development—and over the years in which hundreds of undergraduate and graduate students encountered my lectures on development related topics, the circumstances that I sought to influence—even change—perhaps naively, don't seem to have changed all that much. There seemingly has been little major sustained movement up the "ladder" by Africa—a notion to be reviewed in this volume. To be sure, the continent is more populated and narrative about its predicament has engendered much debate; perhaps so because Africa's development experience has shown considerable ebb and flow. But has there been any appreciable sustained progression? Ebb and flow—yes, but any significantly continuous aggregate upward trend towards the upper rungs of the ladder—surely debatable. Sensitive supporters cry foul at deeply critical essays; critics bemoan "soft" analysis of the "facts." Interestingly, as my emphasis is on a new mode of analysis regarding African development and posits the multinational corporation as key input, the Preface penned for the earlier book draft appears strikingly relevant in present.

As the narrative appeared:

> This book deals with subject matter that in some circles is both sensitive and controversial. Marriages between multinational corporations and Third World hosts have been put to the test in a variety of ways. Some have survived, many have not. For those

---

[1] http://forum.thefreedictionary.com/postst61328_An-invasion-of-armies-can-be-resisted–but-not-an-idea-whose-time-has-come-.aspx

[2] Benjamin F. Bobo, *Corporate and Third World Involvement: A Reciprocal Relationship*, unpublished manuscript (University of California, Los Angeles and University of California, Riverside), 1981, 421 pages.

relationships that have somehow managed to accommodate adversity and compromise, and for those that aspire to play the business enterprise game in a Third World setting, the future is clouded with uncertainty. In many instances the unknowns out weigh the knowns but enterprising entrepreneurs go forth with great zeal encouraged by what appears to be fantastic opportunities for profitmaking and wealth accumulation. Much initiative has been curbed by the reality of the complexity of the international marketplace. But the ultra-strong seemingly thrives on its complexity using it oftentimes as an instrument to shield confidential business operations. Large multinationals are within themselves exceedingly complex organizations. When taken in conjunction with the complexity of the international marketplace they are simply overwhelming.

Host developing countries seemingly find themselves in somewhat of a dilemma: they can't get along with multinationals and they can't get along without them. While this point of view may suffer from exaggeration, it does hold a good deal of truth. The multinational-Third World relationship abounds with examples of conflict and confrontation. At the same time there is considerable evidence of how multinationals aid the economic development process in emerging nations through the provision of direct private foreign investment, management skills and technology. Much of the resources provided by multinationals would not otherwise be available to Third World countries.

The two entities have much to offer both in terms of personal development and in terms of world economic development and stability. Generally speaking, time has been lost and efforts have been wasted owing to the inability of multinationals and host developing countries to establish harmonious working relationships that produce an equitable sharing of wealth derived from the exploitation of business opportunities. A reasonable business operations framework must be established to enable multinationals and Third World countries to engage in business enterprise free of fears of unfair exploitation and threats, of expropriation and nationalization. . . .

The concluding narrative of the earlier exposition, penned as "Reflections," exhibits current relevance as well.

With the growing state of world interdependence in not just business enterprise but in all aspects of human existence, those of us who are sincerely sensitive to the potentialities and realities of emerging conflict between the "haves" and "have nots" sense that the persistent "inequality gap" between advanced and developing nations threatens the future of all mankind. The discussion presented in this book deals with perhaps a small but certainly a very significant issue in world affairs. The multinational corporation is popularly looked upon in many locales around the globe with discernible repugnance. Ostensibly, one quickly concludes that all is not well in the international marketplace.

Where multinationals are concerned, widespread dissension about the benefits of their participation in local enterprise is commonplace in the Third World. Numerous reasons for this have been outlined. In their quest for greater efficiency, larger profits and global visibility, multinationals use their great powers to mobilize capital and deploy it wherever necessary. In the process, the "profit maximization" motive reigns supreme in corporate expansion decisions with seemingly little concern for the aspirations of those who are not fortunate enough to be stockholders.

It is argued that a global factory with a worldwide division of labor is the ultimate in economic efficiency. But what if resource constraints preclude development of a centralized global workplace? Will multinationals then pursue a decentralization strategy, or will they withdraw completely from Third World exploration? It is unlikely that any withdrawal particularly to a significant degree will take place in the near future. The Third World holds an abundance of precious natural resources which multinationals fully intend to exploit. But acts of exploitation must be structured to benefit all parties equitably who have something at stake. Multinationals deserve rewards for risking capital investment and know-how in uncertain environments. Third World people must be compensated for depletion of natural resources. No precise formula to govern the equitable sharing of benefits derived from resource exploitation has been developed. Perhaps equity has been too difficult to define in light of the multinational's profit maximization strategies. When we consider their proclivity for wholly owned ventures in the Third World, we must begin to ask ourselves if equity has any legitimate meaning in globalization strategies. . . .

Without doubt, however, multinational corporations have a special contribution to make to Third World economic development. They provide capital in addition to managerial, technical and marketing skills that cannot be supplied through aid mechanisms or through foreign trade. These factors alone make the MNC exceedingly attractive to emerging nations. But the equity issue looms large in the minds of Third World people and must be satisfactorily addressed by the MNC before any harmonious and prolonged relationship can be established.

The MNC's reluctance and in many cases outright refusal to engage in complete technology transfer, to develop pricing policies that allow host nations to share equitably in revenues, and to adopt accounting practices and procedures that show true and actual revenues and expenses are viewed by the Third World as deliberate efforts to unfairly exploit their resources. Because this behavior is oftentimes quite severe, many host countries fear that a relationship with multinationals will invariably lead to the "development of underdevelopment" . . .

The cumulative and circular causation argument would appear to support this view. As it contends, economic development—if left to assume its own course—is a process of cumulative and circular causation which tends to bestow rewards upon the well-to-do and even to depress the efforts of those lagging behind. Thus a superior-subordinate relationship would develop with industrialized nations reigning supreme and developing countries maintaining an inferior position in the world economic order . . . But a mechanism through which reconciliation can be legitimately attempted if not fully achieved must be found. Host governments remain concerned that a small number of multinational corporations through an intricate system of managerial and financial services, exercise significant control over the . . . development process . . .

Over the years, I have visited and revisited the ideas and ideology entertained in the subject draft, seeking means of expressing the dilemma faced by developing economies in numerous journal articles and formidably in *Rich Country, Poor Country: The Multinational as Change Agent* and *Neo-Liberalism, Interventionism and the Developmental State: Implementing the New Partnership for Africa's Development*.[3] The former, motivated by the

---

[3] See Benjamin F. Bobo, *Rich Country, Poor Country: The Multinational as Change Agent*, Westport, Connecticut: Praeger Publishers, 2005 and Benjamin F. Bobo and Hermann Sintim-Aboagye, eds., *Neo-Liberalism, Interventionism and the Developmental State: Implementing the New Partnership for Africa's Development*, Trenton, New Jersey: Africa World Press; 2012.

earlier book draft, sought to formally situate African development, addressed herein, in the broader frame of Third World theory and practice and to propose multinational corporations as having important input to development policy. Following this effort, the latter directly addresses development policy in the context of the economic malaise confronting African economies through developing a body of knowledge assessing the risks and opportunities associated with implementation of the New Partnership for Africa's Development—a continental policy directive hoped to provide the impetus for sustained economic development.

I now hope to transcend earlier deliberations building upon and extending their fortitudes along with my many years of teaching development-related financial economics to undergraduate and graduate students. In this volume I especially hope to communicate the exciting intellectual challenges of development theory and practice as introduced in the context of the African experience with emphasis on an analytical policy approach to African capacity innovation. Policy analysts, development scholars and practitioners will benefit from an innovative approach to presenting development theory as advanced through an analytical policy framework for capacity innovation as a tool for facilitating sustained modernization in Africa.

The Africa-MNC strategic alliance construct and the policy framework for African capacity innovation articulated herein advises policy analysts on a new and innovative way of thinking about development policy. The seed topic is capacity innovation; it forms the thematic underlay and the strand that ties the book together. Capacity innovation is virtually brought to life through an identification, linkage and elucidation of crucial elements that create a powerful capacity innovation policy structure. The originality of the strategic alliance and policy instrument for capacity innovation will capture and challenge the imagination and foster more critical thinking about and analysis of the African experience. Because of its uniqueness, the book will attract a wide array of supporters and critics; these voices will further its visibility in the marketplace. The critics may be particularly harsh as is to be expected in the face of particularly new and novel ideas. A perspective by Mahbub ul Haq, an author of original discourse on the human development paradigm, is particularly appropriate here: "I have found nothing more fascinating than the birth and evolution of new ideas.... The first stage is characterized by organized resistance. As new ideas begin to challenge the supremacy of the old, all the wrath and scorn is heaped upon the heads of those who have the audacity to think differently ... The second stage can generally be described as widespread and uncritical acceptance of new ideas. At some point, there is a sudden realization that the time for a new idea has arrived and all those who had opposed it thus far hurry over to adopt it as their own. Their advocacy becomes even more passionate than those of the pioneers, and they take great pains to prove that they discovered the idea in the first place . . . It is the third stage which is generally the most rewarding—a critical evaluation of ideas and their practical implementation."[4]

As inscribed in *Rich Country, Poor Country*[5] and subscribed here, the analytical policy perspective is advanced upon an innovative approach to capacity innovation and rests with the multinational corporation as the agent of change with the capacity to motivate policy questions; the moral perspective lies in advising policy analysts on an innovative approach to African capacity innovation and sustained modernization; the intellectual challenge weighs on the capacity of analytical frameworks to adjust traditional thought to accommodate new conceptual approaches; the ethical imperative underscores the need to apply appropriate development analysis to understanding the African experience; and the practical reality impinges the will of policy analysts and practitioners to explore what has yet to be attempted.

---

[4]Mahbub ul Haq, *Reflections on Human Development*, Oxford: Oxford University Press, 1995, 228–229.
[5]Bobo, *Rich Country, Poor Country: The Multinational as Change Agent*, op. cit.

Acknowledgements

I am deeply grateful to Loyola Marymount University as host of this work and to the College of Business Administration for its support. This is the fourth such undertaking that has benefitted enormously from their sponsorship.

I am once again thankful for the tireless support and commitment of Kathe Segall who for the fourth time has done what she has done on the prior three occasions—performed magnificently in putting the manuscript into final form. It has to be rare that an author has the good fortune of having the same superb assistance time and time and time and yet time again. I thank you Kathe yet again for your forbearance and understanding as I have counted on your unwavering attention to my writing projects.

And I once again thank my family for their patience as I imposed on their fortitude in bringing yet another initiative to completion.

Thanks . . .
Thanks . . .
Thanks . . .

List of Exhibits

Exhibit 2.1	Give and Take of a System of Institutions.	17
Exhibit 2.2	Interfacing of Constitution, Rule of Law and Bodies of United States Government.	19
Exhibit 3.1	Stages of Economic Growth.	28
Exhibit 3.2	Ladder of Comparative Advantage.	31
Exhibit 3.3	Core Challenges.	36
Exhibit 3.4	The Mini-states Perspective African Countries by Comparison to G8 Population, GDP and GDP per capita.	39
Exhibit 3.5	The Endgame Perspective Philosophy to Outcome.	50
Exhibit 3.6	Path to Capacity innovation Model of NEPAD Facilitation A Standalone Perspective.	53
Exhibit 4.1	Top Down-Bottom Up Cross Fertilization (Framework of Coordination and Convergence).	72
Exhibit 5.1	Country Capacity ID.	84
Exhibit 5.2	African Countries Core Capacity Attributes.	86
Exhibit 5.3	Lowest Literacy African Countries GDP Per Capita (ppp) World Ranking.	101
Exhibit 5.4	Africa and G8 Relationship of Capital Formation to Gross Domestic Product (2014 – Average).	103
Exhibit 6.1	Africa-MNC Strategic Alliance Operating Structure and Protocol.	119
Exhibit 6.2	Developing a Mindset Operationalizing and Actualizing the Alliance and Balancing Concepts.	123
Exhibit 6.3	African Countries 2015 Corruption Score.	131
Exhibit 7.1	Africa – MNC Strategic Alliance A Policy Framework for African Capacity Innovation.	138
Exhibit 7.2	Characteristics of the Multinational Corporation.	143

Exhibit 7.3	FDI Fund Support by Multinational Corporations.	173
Exhibit 7.4	NEPAD Bank High-Impact Loans/Investments Agricultural, Commercial, Industrial.	175
Exhibit 7.5	Unlocking Core Activities of the Linking Process Africa—MNC Strategic Alliance.	182
Exhibit 8.1	Incorporating FDI Security.	187
Exhibit 8.2	Foreign Direct Investment Security AU's Africa Peace and Security Architecture African Standby Force Mandate.	188
Exhibit 8.3	Strategic Readiness Response Positioning of Security Bases African Standby Force.	190
Exhibit 8.4	FDI Fund Special Fund Provision Peace and Security Council Support by Multinational Corporations.	192
Exhibit 9.1	The Final Exhibit. The Africa-MNC Strategic Alliance Decision-Making Protocol.	197

1

Introduction

There are potentially a number of entry points to erecting stimulus to capacity innovation in Africa; some perhaps more appropriate than others. Celestous Juma, in The New Harvest: Agricultural Innovation in Africa[1] asserts that Africa cannot only feed itself in the span of a generation but contribute to a reduction in global food shortages. Targeting the agriculture sector, Juma outlines an approach to creating what is essentially the capacity among African countries to accomplish these tasks through exploiting technological advances, encouraging entrepreneurship, investing in infrastructure and creating regional markets. This book underscores Juma's perspective but rather than targeting agricultural innovation – as does Juma – the book suggests a more holistic approach to redressing Africa's capacity dilemma by focusing not just on agriculture but the whole of the development spectrum engaging an array of development tools and concepts. The book proposes a framework for African capacity innovation as a path to sustained economic transformation orchestrated through a strategic alliance between African countries and the multinational corporation in which NEPAD (the New Partnership for Africa's Development as a formal operating entity adopted by the African Union to represent African countries) links up and works in alliance with the multinational corporation to address market deficiencies and supply-side constraints as well as strengthen workforce skills and competencies throughout the African continent guided by a detailed capacity innovation scheme through which sustained transformation emerges.

The scheme does not target the agriculture sector, in dissension with Juma's approach, as there is much debate concerning the efficacy of non-market selection of winners and losers[2] particularly as to which types of

---

[1] Celestous Juma, *The New Harvest: Agricultural Innovation in Africa*, Oxford: Oxford University Press, 2011.

[2] Paul R. Krugman, *Strategic Trade Policy and the New International Economics*, MIT Press, 1986.

industries or economic sectors should be targeted and whether such behavior results in the most efficient and cost-effective outcome. Rather, the scheme carries the underlying framing that all economic sectors are within the purview of African economic transformation, not singularly agriculture, and that the African-MNC alliance as proposed in this work encourages an economic structural arrangement whereby private markets emerge and in such fashion as to promote efficiency in resource allocation resulting in market selection of economic drivers. If agriculture, for example, prevails – so be it; if another singular sector or other sectors get the nod, so be this as well. Intervention, as in special interest (government or groups of individuals as in Juma's panel of advisors) sectoral selection can only reduce efficiency. This book shares the conventional wisdom of trade theory that non-market selection of economic drivers, as in the case of Juma's cadre of advisory specialists, interferes with the emergence of private markets and, moreover, the distributional capacity of private markets to accomplish their task in enforcing efficiency. Special interest influence in resource allocation and economic sector selection may not yield the most efficient outcome.

1.1 Direction and Purpose of Book

The direction of the book, and why it should be written, may be better understood by beginning with my reason for writing it. As a researcher and professor of finance and Third World development, I find the development discipline profoundly interesting. My career in teaching and research spans nearly thirty years with a focus on economic strategy, trade policy and Third World development, particularly the African region. African development research is my passion. This book is my third instalment on the topic and forms essentially a "trilogy". I have authored other books and many journal articles but completion of this book capstones a career that has been devoted to understanding the African development dilemma and proposing solution-oriented approaches to sustained economic transformation. This experience leads me to the primary purpose of this volume: to provide scholars and policy analysts an alternative approach to African economic transformation predicated upon the emergence of private markets and competition among private producers. This purpose underlies reasons for writing the book: (1) there is no volume on African capacity innovation comparable to the proposed work, Juma's contribution notwithstanding, (2) the lack of such a volume denies Africa exposure to the innovative approach to African modernity and economic transformation proposed in the book and (3) alternative approaches to capacity innovation afford Africa the opportunity to be more informed and selective in establishing policy directives.

Arguably, inefficient and ineffective capacity innovation policy is Africa's Achilles' heel. In the context of the African state of affairs, this book explores the capacity innovation process akin to the developed world and suggests a remedy particular to the African condition. The remedy foresees the multinational corporation as primary impetus to and facilitator of capacity innovation and an African-MNC alliance as the capacity innovation vehicle. An alliance with multinational corporations through a definitive partnership arrangement has the potential to encourage emergence of private markets and induce long needed sustained capacity innovation.

More precisely to the direction of the book, assisting the New Partnership for Africa's Development is at the very core of my motivation. Many observers of Africa's attempts to undertake capacity innovation strategy capable of putting the continent on a path to sustained transformation held high hopes that NEPAD, adopted by African heads-of-state in 2001, had the potential to accomplish such a feat. Until its adoption, harnessing the productive capacity of the multitude of natural and human resources not only had been elusive but had been viewed with lukewarm expectation of being possible. How to design and orchestrate a successful capacity innovation scheme for Africa is a perennial topic of heated debate; NEPAD is hoped to resolve the debate. But making Africa function effectively as well as efficiently not simply as a group of independent nations but more importantly as a grouping of interdependent nations, believed paramount, is a monumental task. Advantages of private markets, economies of scale and critical mass have not been realized through prior capacity innovation efforts and thus far NEPAD has yet to show significant promise as expected.

The alliance scheme proposed in this work offers NEPAD an alternative way forward. In the scheme, NEPAD signals that African countries are open for business and are organized under a common banner to minimize opportunistic behavior, nationalistic fervor, corrupt practices and disaffection in achieving widespread economic transformation for the common good. Multinationals convey their interest in doing business on the African continent and play the role of change agent, committing resources (FDI) to establishing business enterprise. The ensuing engagement among these entities promotes emergence of private markets whereby resources are efficiently allocated to their highest and best use, economic drivers are established, and goods and services are produced and consumed as orchestrated by the "invisible hand."[3]

1.2 Need and Approach

The focus on NEPAD signals a pledge by African leaders to collectively mount an assault on inadequate capacity innovation necessary to lift African countries out of poverty and underdevelopment. Such effort clearly places the onus on Africa to build necessary capacity to induce sustained modernization. But orchestrating capacity innovation is onerous under the best of circumstances; the structure and composition of African economics is clearly not gifted with the best of circumstances. Africa will have to surmount many challenges that stand at the very heart of the continent's capacity innovation dilemma. How does Africa enable capacity innovation so as to improve African economics to a level sufficient to accomplish sustained modernization?

The book endeavors to articulate an approach to capacity innovation and economic transformation for the African region and in so doing examine the implication and problematic nature of a range of attendant perspectives and capacity innovation frameworks for African modernization including capital theory as a capacity innovation construct, Rostow's stage theory as a capacity innovation process, and the ladder of comparative advantage as an economic structure for capacity innovation. This dialogue will serve as backdrop to articulating a comprehensive approach to capacity innovation for African countries as a collective body through developing a top down-bottom up capacity innovation narrative to enable a more holistic perspective on talents, resources and efforts required for mounting comprehensive capacity innovation, constructing an African capacity ID model to identify and assess the internal resources available to African countries to promote private markets and economic drivers, and proposing a relationship with multinational corporations through an Africa-MNC alliance augmented by a structured foreign direct investment partnership to enable structural transformation. The partnership structure will propose operating features that promote the capacity innovation process. The end result is a strategic capacity innovation policy framework that accommodates Africa's makeup. A final feature of the framework is a proposed array of revisions to the African Union's security protocol for the purpose of protecting the assets and human resources of

---

[3] See Adam Smith, *An Inquiry into the Nature and Causes of The Wealth of Nations (1776), The Franklin Library, Franklin Center, Pennsylvania, A Limited Edition*, 1978, p. 301.

multinational corporations. Attracting foreign direct investment sufficient to induce and support credible capacity innovation and sustained modernization may well depend upon Africa's ability to provide necessary security.

1.3 Objectives

To articulate a path to capacity innovation and ultimately propose a strategic policy framework for doing so, the book:

- Explores capacity innovation perspectives as articulated in the literature and assesses these perspectives in the context of the African predicament regarding emergence of private markets and economic drivers.
- Examines capital theory as a First World development construct in the context of questioning its capacity as an appropriate platform upon which capacity innovation and sustained modernization in Africa may emerge; a system of strategic government institutions is assessed as necessary supporting structure.
- Constructs a top down-bottom up narrative to enable a more holistic perspective on talents, resources and efforts required to mount a comprehensive capacity innovation engagement from the most technologically advanced to the grassroots; the multinational corporation representing the technologically advanced spectrum and the NGO proxying the grassroots.
- Constructs an African capacity ID model to identify and assess the internal resources available to African countries to promote private markets and economic drivers in the context of the range of internal resources required to promote capacity innovation and modernity; the developed country perspective is used as backdrop.
- Assesses impediments to African capacity innovation with a view towards establishing a perspective on the strategic fixes that a capacity innovation initiative must address in the context of promoting private markets.
- Proposes a structured alliance between Africa and the multinational corporation to enable efficient attraction, allocation and implementation of foreign direct investment.
- Proposes opportunistic balancing to discourage opportunistic behavior by parties of the alliance.
- Articulates a strategic capacity innovation framework to promote emergence of private markets, competition among private producers, and market selection of economic drivers.

- Enunciates a strategic framework to accommodate Africa's makeup; supply – and demand-side features are incorporated along with implementation strategy.
- Proposes a foreign direct investment protection scheme to provide necessary security for facilities and human resources.

1.4 Undertaking the Objectives

The objectives pose significant challenge to intellectual resolve in light of the daunting prospect of redressing African capacity shortcomings. Taking up this challenge, Chapter 1 conveys the purpose of the initiative: to provide scholars and policy analysts an alternative approach to African economic transformation and in so doing articulate a framework for African capacity innovation as a path to sustained modernization orchestrated through a strategic alliance between African countries and the multinational corporation in which NEPAD links up and works in partnership with multinationals to address market deficiencies and supply-side constraints as well as strengthen workforce skills and competencies throughout the African continent guided by a detailed capacity innovation scheme through which private markets and economic drivers emerge. Chapter 1 illuminates the character and coverage of each chapter in the book.

Chapter 2 assesses capital theory as a capacity innovation construct and challenges its oversight in clearly advising that a system of strong institutions is essential to the smooth operation of a capitalist structure and moreover in addressing corruption. The chapter examines capital theory as a First World capacity innovation construct in the context of questioning its potential as an appropriate platform upon which capacity innovation and sustained modernization in Africa may emerge; a system of strategic government institutions is assessed as necessary supporting structure.

Chapter 3 explores capacity innovation perspectives as articulated in the literature and assesses these perspectives in the context of the African predicament. The chapter assesses Rostow's stage theory and the ladder of comparative advantage as capacity innovation constructs and uses these to prepare an analytical backdrop for devising a standalone conceptual policy framework for African capacity innovation. This is the initial step to articulating a holistic approach to capacity innovation in Africa.

Chapter 4 constructs a top down-bottom up narrative to enable a more holistic perspective on talents, resources and efforts required to mount a

1.4 Undertaking the Objectives 7

comprehensive capacity innovation engagement from the most technologically advanced to the grassroots; the multinational corporation representing the technologically advanced spectrum and the NGO serving the grassroots. Chapter 5 constructs an African capacity ID model to identify and assess the internal resources available to African countries to promote private markets and economic drivers in the context of the range of internal resources required to promote capacity innovation and modernity; the developed country perspective is used as backdrop.

Chapter 6 articulates an alliance between NEPAD and the multinational corporation as a formal and efficient means of achieving sustained capacity innovation through addressing market deficiencies and supply-side constraints as well as strengthening workforce skills and competencies throughout the African continent; assesses impediments to African capacity innovation with a view towards establishing a perspective on the strategic fixes that a capacity innovation initiative must address in the context of promoting private markets; proposes a structured alliance between Africa and the multinational corporation to enable efficient attraction, allocation and implementation of foreign direct investment; and articulates the notion of opportunistic balancing to mitigate the power of multinationals and African countries to behave opportunistically.

Chapter 7 accomplishes the ultimate purpose of the book in presenting the capacity innovation policy framework undertaken in this volume through articulating a scheme in which capacity innovation is virtually brought to life through an identification, linkage and elucidation of crucial elements that create a powerful capacity innovation structure using an alliance between NEPAD and the multinational corporation to bring about reality. The chapter articulates a strategic capacity innovation framework to promote emergence of private markets, competition among private producers, and market selection of economic drivers engaging an array of development tools and concepts including government institutions and institutional capacity, World Bank, IMF and NGO input, foreign direct investment, comparative advantage, regional integration, economic complementarity, price-maker signaling, intra-continental communication, infrastructure development, critical mass, trade misinvoicing and transfer pricing, foreign direct investment fund and a strategic banking scheme, and a security apparatus to protect foreign direct investment. A strategic capacity innovation framework is enunciated expressly devised and structured to accommodate Africa's makeup; supply – and demand-side features are incorporated along with implementation strategy.

8 *Introduction*

Chapter 8 undertakes the final feature of the volume through a proposed array of revisions to the African Union's security protocol for the purpose of protecting the assets and human resources of multinational corporations. Attracting foreign direct investment sufficient to induce and support private markets may well depend upon Africa's ability to provide necessary security. To this task, a foreign direct investment protection scheme is proposed to provide security for facilities and human resources; the African Union's peace and security architecture and the African Standby Force provide the security apparatus.

In closing, Chapter 9 offers an epilogue; taking a look back at the accomplishments of the book in light of its objective and situates it in the larger scheme of Africa's struggle for sustained economic transformation. Achieving this feat is ultimately, the book affirms, facilitated by the Africa-MNC strategic alliance decision-making protocol – the interactive decision-making process that enables the alliance relationship.

2
Capital Theory and Development: Has Africa Been Overlooked?

The World Commission on Environment and Development defines sustainable development (capacity innovation) as "development that meets the needs of the present without compromising the ability of future generations to meet their own needs."[1] When asked why Africa has been resistant to sustainable capacity innovation, a frequent retort points to government corruption: misfeasance, malfeasance, intervention in resource allocation to accommodate special interest, political action to select winners and losers in lieu of private markets, pillaging the public treasury, and the like. In a study of corruption and economic growth, Pak Hung Mo found that a 1% increase in the level of corruption reduces the growth rate by about 0.72%, noting that this outcome derives to a large extent from political instability, accounting for about 53% of the total effect. Mo confides that corruption tends to more pejoratively impact innovative activities. This is manifested in a number of ways: innovators require government-supplied goods, such as permits and import quotas, more so than established producers – the demand for which is significant and inelastic thereby providing ready targets of corrupt practices; typically having no established lobbies and political connections, innovators are especially subject to bribes and expropriations; often credit-constrained, innovators are unable to pay bribes thereby leading to reduced private investment and human capital as innovative activities are foregone; and, ultimately, reduced private investment is unfavorable to capacity innovation.[2]

---

[1]David I. Stern, "The Capital Theory Approach to Sustainability: A Critical Appraisal," *Journal of Economic Issues*, Vol. XXXI, No. 1, March 1997, p. 146. There is ongoing debate on the definition of sustainable development; however neoclassical economists generally define it as non-declining average human welfare over time. Ibid., p. 147.

[2]Pak Hung Mo, "Corruption and Economic Growth," *Journal of Comparative Economics*, 29, pp. 66–79, March 2001. See also, Paolo Mauro, "Corruption and Growth," *Quarterly Journal of Economics*, 110, 3:681–712, August 1995.

A study by Kar and LeBlanc provides further insight on this matter. Investigating illicit financial flows from developing countries, they assert that illicit financial outflows are the most devastating economic issue impacting poor countries. Interestingly, their findings showed scant evidence that macroeconomic conditions give rise to illicit flows but revealed that government corruption and export misinvoicing (to be addressed in Chapter 7) are notable drivers. On the matter of government corruption, Kar and LeBlanc noted that corruption is related to the state of overall governance in the country. The extent of corruption in specific countries is captured by The World Bank governance indicators by measuring the world percentile rank of the countries' corruption indicators; a decline in the percentile rank denotes greater corruption relative to other countries and vice versa.[3]

Clearly government corruption is not simply an African phenomenon, it is rather omnipresent. Developed as well as developing economies are somehow afflicted with it.[4] The intensity of the affliction notwithstanding, its wanton power raises more than idle curiosity particularly in the context of African capacity innovation. Reconciling African development incapacity may be, in many ways, a reflection of the limitations of capital theory, and perhaps even more, capitalist institutions. Development literature is replete with high-powered mathematics, grandiose schemes and intellectual verbiage on the subject touting the wherefores of course corrections essential to a fix. Failure in accomplishment potentially underscores not knowing how to fix the problem or not caring how to do so. Both likely hold the key to understanding the paucity in African capacity innovation. Capital theory in all its grandeur encapsulating price system and resource allocation, saving and investment, factors of production, net present values, rents, interest rates and the entire scope of time value of money has not empowered African capacity innovation so as to overcome the shackles of underdevelopment. Though it is credited with giving rise to the First World construct, it has been rather ineffective in resolving the Third World construct, a resolve that begs the question – is there a fix, particularly for Africa.

I don't by any direction of the moral compass imagine that I possess the skill to redress the many shortcomings of economic scholarship surrounding

---

[3] Kar and LeBlanc found that US$946.7 billion in illicit outflows left the developing world in 2011, compared with US$832.4 billion in 2010. See Dev Kar and Brian LeBlanc, "Illicit Financial Flows from Developing Countries: 2002–2011," Global Financial Integrity, December 2013.

[4] See Corruption Perception Index 2015, Transparency International, http://www.transparency.org/research/cpi

capital theory but my energy so virulently flows in the direction of government corruption oversight that I feel compelled to at least attempt in my limited capacity to set the record straight, so to speak. For the sake of writer's privilege, I disassociate the mounds of literature on capital theory, however intellectually sound they may be, so that I may undertake my objective with minimum confluence or distraction.

2.1 Capital Theory and Government Corruption

So then, I feel compelled to ask, should capital theory accommodate government corruption? This question is particularly relevant since, theoretically, the literature reaches no agreement about the effect of corruption on economic growth.[5] In *Capital Theory and the Rate of Return*, Robert Solow sheds light on the matter by asking perhaps the more basic question: what is the proper scope of capital theory?[6] He reacts to the question from essentially three perspectives. The theory of capital is connected with the microeconomic theory of capitalist structure in the mode of competitive pricing and resource allocation. It may be further viewed as the explication of the causes and consequences of saving and investment. And, capital theory in ways, the third perspective, subsumes notions of price and resource allocation in combine with efforts to answer questions about the consequences and causes of saving and investment decisions. Beyond these, a fourth perspective may be in order – more so as a reminder: investment decisions are not unique to individuals, households and firms, but include government as well. Government investment in infrastructure and the human paradigm[7] supports the capitalist structure. Hence resource allocation as prescribed by capital theory is impacted by government investment decisions.

Resource allocation therefore offers a particular platform from which government corruption may be examined. Government corruption is much like, for example, corporate corruption – a violation of trust. Trust is implied in every walk of life whether, for example, in buying goods and services, selling goods and services, investing in the production of goods and services, providing transfer payments for the purchase of goods and services, and the

---

[5]See Pak Hung Mo, p. 66.

[6]Robert M. Solow, *Capital Theory and the Rate of Return*, Chicago: Rand McNally & Company, 1964, p. 14.

[7]Mahbub ul Haq, *Reflections on Human Development*, Oxford: Oxford University Press, 1995.

like. The acts involve decisions that impact the allocation of resources. One trusts another not to cheat or abuse the privileges society offers. Even more, a system has manifested in society that orchestrates behavior in a large way so as to minimize the need for distrust or fear of mistrust. Such a system is explicated in the treatise of capital theory.

Capital theory quite simply is about the interaction of households and firms[8] as actors or role players in a game of demand and supply which creates a marketplace or a free market for goods and services and implies an underlying model of capacity innovation. In this game, the households as well as the firms play dual roles: at times the households are workers in the firms that produce goods and services to satisfy demand for them in the marketplace, and in role switching they are consumers in the marketplace where bidding for goods and services takes place – creating the front end (or perhaps back) of an invisible hand-like market function that responds to bidder desires without allowing any single bidder (consumer) to control market function. In a similar scenario, firms at times are suppliers of goods and services through employing workers to produce them and in role switching they are also consumers in the marketplace where bidding for goods and services takes place – creating the back end (or perhaps front) of an invisible hand-like market function that responds to their desires without allowing any single bidder (producer) to control market function.[9] This seemingly magical process creates competition for goods and services which establishes prices of these products at which consumers (households) claim their bids and suppliers (firms) satisfy their bids. This process of competition and pricing among the actors (role players) leads to an efficient allocation of goods and services, or reasonably so. The end result is a manifest price system and resource allocation – an arrangement that gives rise to a number of consequences – among them winners and losers in investment transactions, comparative advantage, economic take-off and ultimately sustained economic capacity innovation (discussed in Chapter 3).

In the marketplace, the consumers are identified by various descriptions: poor and rich; lower, middle and upper class; haves and have-nots; small, medium and large firms; government/government agencies; taxpayers; investors; and the like. The interaction of these entities in the marketplace presents a complex structure of desires giving rise to, from time to time,

---

[8] Speaking on capital theory and a state of equilibrium, Bliss notes that "there are two kinds of actor, the household and the firm." C.J. Bliss, *Capital Theory and the Distribution of Income*, New York: American Elsevier Publishing Company, Inc., 1975, p. 20.

[9] The general marketplace/free market is characterized by 'dual' marketplaces whereby households typically bid for finished goods and firms typically bid for unfinished goods.

2.1 Capital Theory and Government Corruption

advantage-taking or opportunistic behavior. But the power of the marketplace as wielded by the invisible hand is perceived as having the capacity to keep wrongdoers in check. This therefore begs the question: if the invisible hand has the power to regulate the behavior of wrongdoers, so to speak, why is government corruption such a pejoratively imposing force particularly in the African context where government corruption is credited in a large way with impeding capacity innovation. Government is but one player in the marketplace, albeit significant. But government intervention in private/free markets to select what it believes will be "winning" investments should be accountable to the market mechanism and indeed is so. In the instance in which government bypassed the market selection process, as done for Solyndra LLC,[10] the invisible hand reacted with retribution much in the same manner as it did in the cases of a firm that cheated – Enron Corporation,[11] and an individual who misbehaved – Bernard Madoff.[12] All were dealt terminating blows.

However, while we pay homage to the power of the invisible hand, it is important to point out, if not remind us, that it does not singlehandedly impose retribution. In reality, the invisible hand is backed by the full power of supporting government institutions – particularly legislative and judicial. These combine to bring abusers to a day of reckoning. But is capital theory necessarily clear on this reckoning process especially regarding the government element? Does it articulate well enough or clearly call out specification for government input as it does for investment, time value of money, or maximizing net present value,[13] or clearly explicate the exigencies of government

---

[10] Many observers regard the Solyndra LLC $535 million U.S. government subsidy as a corrupt act since it was a non-market decision involving taxpayer funds in support of a for-profit firm that resulted in failure. In ways, it raises the specter of government's inability to pick "winners and losers."

[11] Enron Corporation an energy supplier was one of the largest bankruptcies in U.S. history. The U.S. Securities and Exchange Commission investigation of the company uncovered a number of violations including concealed expenditure and losses, overstated profits, tax evasion, evidence destruction, improper management compensation and violation of the agency relationship between managers and shareholders. Enron filed for bankruptcy on December 2, 2001.

[12] Bernard Madoff, chairman of a Wall Street securities investment firm, operated the largest Ponzi scheme in U.S. history involving investment fraud of nearly $65 billion. In violation of U.S. securities law Madoff paid returns to existing investors from new capital received from new investors rather than from profits earned from invested capital. In 2009 he was sentenced to a prison term of 150 years.

[13] Stern, op cit, p. 147.

corruption upon private or free market functions? To its credit, however, in *The Capital Theory Approach to Sustainability: A Critical Appraisal*, Stern notes that models explicating the capital theory approach to sustainable development "assume that sustainability is technologically feasible and examine under what institutional conditions sustainable development actually will be achieved."[14] He further confides that "Institutional Capital includes the institutions and knowledge necessary for the organization and reproduction of the economy."[15] Still, does capital theory clearly advise that a system of strong institutions is essential to the smooth operation of a capital structure? This may be presumed in a system (developed economy) endowed with efficient institutions and a free market structure,[16] but for a system overwhelmed with acts of government corruption, as in the African case, clarity is no less than obligatory. There should be no blindness on this matter by African government or the African people. Africans must be alerted to the inescapable reality that capital theory, that is – the capitalist institution – does not and cannot stand alone as a tool of redress for government corruption. A supporting cast – government institutions – is requisite.

2.1.1 Government Institutions

While capitalists extol the virtues of private markets and profess a theoretical construct that rationalizes confidence in free markets and market forces as means of promoting efficient capacity innovation, there is more to be said about how these means are propagated and sustained. To be sure, private markets do about as well as can be expected in promoting efficiency,[17] but free and indeed active markets are not immune to special interest influences thereby requiring the assistance of government institutions in roles of oversight and corrective action to improve market performance from time to time.[18] Milton Friedman,

---

[14] Ibid., pp. 148–149.

[15] Ibid., p.150.

[16] Since developed economies didn't start out in a developed state and have encountered much government corruption, perhaps capital theorists should be less presumptive.

[17] See James A. Brander, "Rationales for Strategic Trade and Industrial Policy," in Paul R. Krugman, (Ed.), *Strategic Trade Policy and the New International Economics*, Cambridge, Massachusetts: MIT Press, 1986, p. 24.

[18] See Karl Botchway and Jamee Moudud, "Neo-Liberalism and the Developmental State: Consideration for the New Partnership for Africa's Development," in Benjamin F. Bobo and Hermann Sintim-Aboagye, (Eds.), *Neo-Liberalism, Interventionism and the Developmental State: Implementing the New Partnership for Africa's Development*, Trenton, New Jersey: Africa World Press; 2012, pp. 40–41.

in *Capitalism and Freedom*, stresses that "The existence of a free market does not of course eliminate the need for government. On the contrary, government is essential both as a forum for determining the "rules of the game" and as an umpire to interpret and enforce the rules decided on. What the market does is to reduce greatly the range of issues that must be decided through political means, and thereby to minimize the extent to which government need participate directly in the game. The characteristic feature of action through political channels is that it tends to require or enforce substantial conformity. The great advantage of the market, on the other hand, is that it permits wide diversity. It is, in political terms, a system of proportional representation. Each man can vote, as it were, for the color of tie he wants and get it; he does not have to see what color the majority wants and then, if he is in the minority, submit."[19] Still again, does capitalist treatise make clear, indeed poignantly comprehensible, that a system of strong institutions must accompany, if not undergird, a successful capitalist structure? Is not a successful free market only possible when operating at least in tandem with strong institutions?

Despite the very impressive explication of capital theory, its superior touting can only prove itself with the aid of a good institutional mapping, in short – good governance. An institutional mapping characterized by interfacing institutions is essential for an efficiently functioning market. In an overarching manner, rule of law – a necessary accompanying institution – assures the integrity of the complex system of institutions themselves. While the market has ways to punish some violators, it does not reproach all. In the case of firms that cheat, say, by ignoring sound accounting and capital budgeting practices or investment protocol, firm failure as punishment may not directly hold accountable managers who devise the cheating; the failures of Enron Corporation and Bernard L. Madoff Investment Securities LLC so attest. Likewise, in the case of corrupt officials who demand kickbacks or accept bribes, say, for endorsing infrastructure projects, or government officials who attempt to pick "winners" as favors to campaign donors, the most willing or deserving participants may not offer project proposals or receive support for projects that are the most efficient solutions for the country, hence a resulting punishment of the country but not the corrupt officials. So how do cheating managers and corrupt officials atone for their indiscretions? Invariably, the institutional rule of law along with supporting institutions is called upon to punish the wrongdoers.

---

[19]Milton Friedman, *Capitalism and Freedom*, Chicago: University of Chicago Press, 1962, (p. 21).

In the former which invariably involves violation of the agency relationship between managers and shareholders[20] and loss of shareholder value, firm failure is a function of the market mechanism. But the interplay of the institutional mechanism is necessary to indict and punish managers under the rule of law, which may not only involve the courts but the penal system as well. In the latter, private firms having developed capacity through market competition to more efficiently engage large government projects may be denied such opportunity owing to corrupt government practices. Selection of firms not best qualified to do the job or not groomed by the market invariably places public funds to less than highest and best use, hence in this instance the market's way of punishing ignoring of its power to prepare competitive or best qualified firms. Here as well however, the wrongdoers – corrupt officials – would escape accountability without the interplay of the institutional mechanism to indict and punish.

2.1.2 Give-and-Take

In Solow's book, the silence in dialogue on government intervention – particularly the very dangerous and exceedingly interdicting 'government corruption' – appears to be rationalized by imbedding the omission in the question of "scope" and thereby setting the boundaries of the neo-classical theory of capital. But does what may be simply a rather convenient definition of "scope" absolve capital theory of any responsibility for government action, certainly directly?[21] Certainly we make allowances for the nature of theory, but a paradigm that is the expression of the philosophy and behavior of the most powerful, productive and plentiful economic system on the planet should at the very least clearly acknowledge the fusing of the capitalist and political structures. There is an indisputable give-and-take between the two structures that renders success on either side dependent upon the other.

Exhibit 2.1 depicts the give-and-take of a system of institutions in which the free market system is the core of an institutional mapping. This in reality portrays the actual workings of capital theory/the capitalist model as a construct bounded by the capitalist institution functioning in concert with supporting institutions. As a practical matter, the capitalist structure

---

[20]The agency relationship between manager and shareholder – the principal-agent problem – holds that the manager as agent of the shareholder (principal) acts in the best interest of the shareholder not in his/her self-interest.

[21]Solow, p. 14 (paraphrase the 2nd and 3rd paragraphs).

Exhibit 2.1 Give and Take of a System of Institutions.

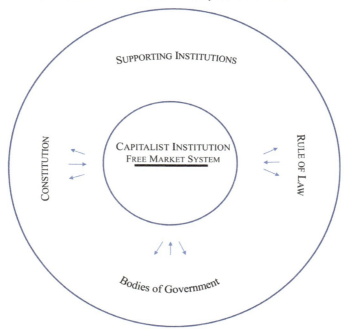

empowers the supporting institutions and the supporting institutions empower the capitalist structure in a give-and-take arrangement. The free market system as an institutional construct exemplified by the workings of the invisible hand organizes production and allocates resources. This system is the tour de force in capitalism-led development. Supporting institutions characterized by a constitution, rule of law and bodies of government provide auxiliary capacity to the capitalist structure exemplifying the role of the state in development.

To further elucidate, Exhibit 2.1 prescribes a complete institutional mapping characterized by two features – free enterprise and governance. The former is orchestrated by the invisible hand – a process earlier outlined. The latter prescribes a government apparatus empowered by governmental decree and guided by a government philosophy or set of guiding principles (a constitution); the rule of law to assure observance of the guiding principles; and bodies of government to facilitate the government apparatus. At the very core of the institutional mapping is how much power to allow the central government. This is particularly important in assuring that the free market feature is not compromised by special interest.[22]

2.1.3 The Government Apparatus: Governmental Decree

Outside the workings of the market mechanism, which plays an interfacing role in adjudicating improper behavior, the power to indict and punish is clearly held by governmental decree and vested in governance. This necessary power/authority is set forth through the Constitution, rule of law and bodies of government. Through these comes the very essence of a "system" of government. In the Western context particularly, the "system" underlies a mindset characteristically "of the people, by the people and for the people" for which rules of behavior are established; the power to indict and punish flows through this mechanism. The U.S. model, as portrayed in Exhibit 2.2, serves to illustrate the "system" but similarities may be found in the institutional makeup of other democracies.[23] The interfacing of the Constitution, rule of law and bodies of government illustrates the broad character of the U.S. institutional mapping upon which "law" is the power principle, hence establishing the power to indict and punish. This power may be gleaned from important passages in the institutional prescripts.

2.1.4 Constitution

Article Six of the U.S. Constitution provides that the Constitution and the laws of the United States including all treaties shall be "the supreme law of the land." It further decrees that the "Judges in every State shall be bound thereby." From these supreme edicts emerge a number of passages underlying the power to indict and punish including the power ... "to make all laws which shall be necessary" ... as well as:

---

[22] For a recitation on institutional mapping, see Benjamin F. Bobo (2012). Implementing State Interventionist Development: The Role of the Multinational Corporation. Chapter 9, in Benjamin F. Bobo and Hermann Sintim-Aboagye, (Eds.), *Neo-Liberalism, Interventionism and the Developmental State: Implementing the New Partnership for Africa's Development*, Trenton, New Jersey: Africa World Press. (pp. 220–223); and Hilton L. Root, (1996), *Small Countries, Big Lessons: Governance and the Rise of East Asia* (see Chapter 10, "The Search for Good Governance"), New York: Oxford University Press.

[23] Constitution of the United States, http://www.senate.gov/civics/constitution/item/constitution.htm See also Similarities, Differences Between the U.S. and Other Major Democracies, http://iipdigital.usembassy.gov/st/english/publication/2008/06/20080628202248eaifas0.9543421.html

- ... the privilege of the Writ of Habeas Corpus shall not be suspended ...
- ... no state shall ... pass ... ex post facto law ...
- ... the judicial power of the United States shall be vested in one supreme court and in ... inferior courts ...
- ... the judicial power shall extend to all cases, in law ...
- ... the trial of all crimes ... shall be by jury ...

These directives clearly convey the resolve of the Constitution to establish the framework for the rule of law. The notion of making all laws which shall be necessary carries the weight of absoluteness in which the rule of law extends to all individuals. Violation of the rule of law – for example, commission of a crime as judged by the rule of law – presupposes a reckoning process with the power to indict and punish so vested. This emerges in part from the directive that the trial of all crimes shall be by jury.

Exhibit 2.2 Interfacing of Constitution, Rule of Law and Bodies of United States Government.

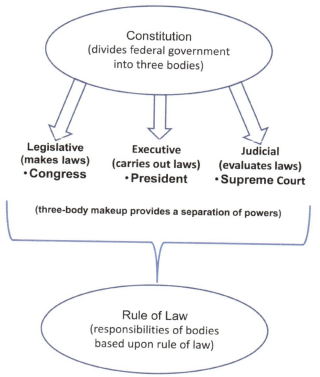

2.1.5 Rule of Law

Conceptually the "rule of law" conveys the emplacement of standards of conduct upon which a society operates. A broad definitional representation would be: consensus or agreement among people about how they will conduct themselves in society. The American Bar Association's World Justice Project has proposed a working definition that comprises four principles: "a system of self-government in which all persons, including the government, are accountable under the law; a system based on fair, publicized, broadly understood and stable laws; a fair, robust, and accessible legal process in which rights and responsibilities based in law are evenly enforced; and, diverse, competent, and independent lawyers and judges."[24] Though there may be varying impressions of the rule of law, these edicts convey the essential idea. It is through these edicts that the power to indict and punish transpires.

2.1.6 Bodies of Government

The U.S. Constitution divides federal government into three bodies – Legislative, Executive and Judicial. The three-body makeup provides a separation of powers. Interestingly, the purpose of each body is connected to "law" as the responsibilities of the bodies are based upon the rule of law. The Legislative body (Congress) makes laws, the Executive body (President) carries out laws and the Judicial body (Supreme Court) evaluates laws. Law is clearly the power principle in this arrangement; such power emanating from the Constitution and the laws of the United States as the supreme law of the land. These responsibilities impact the entirety of the institutional mapping – the free enterprise system and the governance system as well. The rule of law assures that the invisible hand essentially operates in an unfettered manner[25] and that the government apparatus observes the guiding principles.

2.1.7 So – Has Africa Been Overlooked?

Having said all that's been said on the capital theory matter, an epilogue of sorts is ventured. At critical junctures, the capitalist structure as it imbues free market capitalism – or fails to do so – benefits from the helping hand of the political structure. There are clear examples of the failing of the market mechanism: the Great Depression, Asian debacle, Subprime fiasco,

---

[24]"What is the Rule of Law," American Bar Association, 2007.
[25]There are exceptions of course: the Federal Reserve manipulating interest rates, federal government approval of corporate mergers, etc.

2.1 Capital Theory and Government Corruption 21

etc., and government intervention that saved the day – the New Deal,[26] U.S. Federal Reserve rescue[27] and the Troubled Asset Relief Program (TARP)[28] bailout. These speak nothing of the many instances of government corruption – officials taking 'kickbacks' for engineering favorable government policy as well as directing government funds in the best interest of special parties. Such government behavior invariably impacts the 'investment' parameter in the capitalist structure.

Capitalism is not a panacea as conventional free market wisdom ofttimes seemingly purports; 'neo-liberal economic policy' is given preeminence over the 'role of the state' – perhaps blindly so at times. But does the time-to-time fusing of 'market operations' with 'government will' and vice-versa prompt theorists to suggest or propose some sort of purity in capital theory? As fully as intellectual capacity allows, should capital theory not recognize its shortcomings as well as virtues? Overlooking government corruption or excluding it from capital theory dialogue is perhaps an oversight or simply that it is not clearly explicated. Clearly all aspects of capital structure or impacts thereon are of vital concern, the difficulty of representing them in a single theory notwithstanding. And for reasons owing to the multiplicity of aspects or impacts, the theory of capital has been and remains a subject of controversy. But in the African case in which an excluded impact is so pejoratively profound as to override capitalist or free market will, government corruption needs to be formally addressed by capital theorists in the most assertive manner despite the absence of agreement or empirical support. Ignoring this leads to raised expectations which does disservice not only to the tenets of the theory but to market 'players' who aspire to the principles upon which capital theory is propagated. Or worse, disadvantaged observers of the outcome of capitalist production and free market enterprise may become disillusioned

---

[26] See Eric Rauchway (2008). *The Great Depression & The New Deal*, New York: Oxford University Press.

[27] Recovery from the 1997–98 Asian debacle, a financial crisis triggered by easy borrowing and rapid capital flows to the region, was in part aided by the U.S. Federal Reserve by reducing interest rates in the United States leading to a flow of funds back to the Asian region. See Jomo Kwame Sundaram, "What did we really learn from the 1997–98 Asian debacle?" in Bhumika Muchhala (Ed.), *Ten Years After: Revisiting the Asian Financial Crisis*, Washington, DC: Woodrow Wilson International Center for Scholars 2007 (pp. 21 and 26).

[28] To respond to the U.S. subprime mortgage crisis, in 2008 the federal government instituted a program to purchase "acid" assets from financial institutions to strengthen the banking sector. The crisis began in late 2007 resulting from a massive drop in home values leading to mortgage delinquencies and foreclosures creating a national banking emergency and ultimately triggering a recession.

that a capitalist structure may not bear their resolve. Avoiding the latter may be imperative as disadvantage is no stranger to Africa. Africans must clearly understand that sustained capacity innovation will require a capitalist structure in which an efficient market mechanism is at work and a definitive institutional mapping in which efficient government institutions are in play. Resolution of African economic incapacity will depend upon it.

Capitalism is a worthy system, but one shouldn't stand in waiting to be fed by it; having a full scope of how it works notwithstanding. It dispenses fish and will even teach one how to fish, but it requires particular capacities: ready, willing and able, and a sense of collective self-reliance to induce forces of change and fuel flames of opportunity. One must be ready to engage the unknown; willing to take risks that may obtain winners as well as losers; able to afford the costs of pursuit be they emotional, physical or pecuniary; and exhibit solidarity in ways that enable collective forces of change and induce sustained progress of the masses. The path of progress is long and challenging. In furthering understanding of the path of progress particularly with eyes on Africa, the following chapter explores the nature of the path and the essence of the challenges.

3

Towards an Analytical Policy Framework for Assessing Capacity Innovation: A Standalone African Perspective*

> Is a country
> poor because it is poor
> poor because of poor policies[1]
> poor because of poor policy implementation?

Africa is in dire need of taking control of its future capacity innovation process lest it renders such control, as to the West in the past, and now to the very apparent and indisputable intentions and exploits of China. To be sure, Africa would be ill advised to reject the potential benefits that China may offer. But buyer beware. Chairman Mao inked a blueprint for African venturism when proposing that "What the imperialists fear most is the awakening of the Asia, Africa and Latin American peoples ... We should unite and drive the US imperialism from Asia, Africa and Latin America back to where it came from."[2] Despite the pursuer, interest in Africa stems from the same motives: natural resource exploitation and political hegemony. But these are not necessary evils. No doubt this perspective conjures a philosophical divide in the development community. Importantly however is that what has given the foreigner's quest for African presence a bad flavor, be it resource or hegemony motivated, is that Africa has not controlled that quest and channeled it into strategically productive outcomes in the best interest of the continent. This shortcoming has potential resolve in the New

---

*By permission of *Journal of Global South Studies* (formerly the *Journal of Third World Studies*)

[1] Gerald M. Meier, *Leading Issues in Economic Development*, New York: Oxford University Press, 1989, p. v.

[2] Cited in Kehbuma Langmia, "The Secret Weapon of Globalization: China's Activities in Sub-Saharan Africa," *Journal of Third World Studies*, Volume XXVIII, No. 2, Fall, 2011, p. 46. See also Domingos Jardo Muckalia, "Africa and China's Strategic Partnership," *Africa Security Review*, 13, No.1 (2004): 5–11, 6.

Partnership for Africa's Development, but only if the partnership overcomes fractionalization. Some fifty-five African countries must muster the collective will to mount sufficient capacity to avoid the economic miscues of the past. If efficiently orchestrated, NEPAD can attenuate foreign political agenda as well.

It would be novel to say that outsiders resort to divide and conquer tactics to advance economic and political objectives on the continent. But African countries have displayed such little unity that such tactics have been little necessary. Arguably Africa has been a continent divided. NEPAD offers the opportunity not only to meld a collective African body but to put forth a unified front against external forces and ultimately to orchestrate a winning development strategy. This must come from what has been acutely lacking on the continent – a successful capacity innovation initiative. And surely, it must be no less profound than that imposed on the continent by outsiders to advance the priorities of outsiders.

Clearly, any agenda that prioritizes the needs of Africans as subordinate matters is not in the best interest of Africans. The Westerners built capacity in Africa for themselves; now the Easterners are engaged similarly. All along, the Africans have stood on the sidelines, so to speak, watching the show. NEPAD proposes to change the equation. To do so, it must create capacity whereby African priorities are the primary units of input, African market forces are the primary orchestrators of production and African people are the primary beneficiaries of output. This exemplifies the traditional input-output construct which stands as the core economic growth model that has led to such historically profound modernity in Western society. The change element is that African development is the objective; a diversion from the Western tradition. It may be all too apparent; Africa may need to create capacity a new fashioned way—do as they do, not as they say.

But doing as they do may be a task of immense proportions. However, many observers of Africa's attempts to undertake capacity innovation strategy capable of putting the continent on a path to sustained growth hold high expectations that NEPAD has the potential to accomplish such a feat even as a standalone initiative, that is, without the control and power of FDI and/or foreign aid. Until NEPAD's adoption, harnessing the productive capacity of the multitude of natural and human resources required to accomplish such a task not only had been elusive but had been viewed with lukewarm expectation of being possible. How to design and orchestrate a successful capacity innovation plan for Africa is a perennial topic of heated debate; NEPAD is hoped to resolve the debate. But making Africa function effectively as well as efficiently not simply as a group of independent nations but more importantly as a grouping of interdependent nations, believed paramount, is a monumental task. Advantages of economies of scale and critical mass have not been realized through prior efforts and thus far NEPAD has yet to show seriously measurable promise. This is never more apparent than the impression one receives when appraising the condition of the continent.

3.1 Africa's Predicament

Africa is quite literally in the throes of persistent capacity innovation crises.[3] Efforts to put it on a sustainable course of capacity innovation have had mixed and ofttimes largely disappointing results.[4] Numerous ill-conceived capacity innovation experiments over the past 25 years so attest.[5] The continent's latest capacity innovation scheme, NEPAD, perceived to be a partnership between African leaders and G8 countries (US, UK, Canada, Japan, Germany, France, Italy, and Russia), is now a work in progress, so to speak. Africa has high hopes that NEPAD will eventually provide the direction and impetus for sustained economic development so long sought but it has yet to live up to expectations.[6] Persistent socio-economic conditions across the continent suggest that Africa faces a formidable challenge.

Any initiative conceived to confront Africa's socio-economic conditions must clearly recognize the exigencies of the African environment. Some 400 million people, about half the continent's population, live on less than US $1.25 per day.[7] Africa has the highest rates of child mortality; 1 child in 8 dies before

---

[3]Maureen Kihika, "Development or Underdevelopment: The Case of Non-Governmental Organizations in Neoliberal Sub-Saharan Africa," *Journal of Alternative Perspectives in the Social Sciences*, Vol. 1, Issue 3, Dec. 2009, pp. 783–79.

[4]See Johannes Tsheola, "Global 'openness' and trade regionalism of the New Partnership for Africa's Development," *South African Geographical Journal*, Vol. 92, Issue 1, 2010, pp. 45–62 and Min Tang and Dwayne Woods, "The Exogenous Effect of Geography on Economic Development: The Case of Sub-Saharan Africa" African & Asian Studies, Vol. 7, Issue 2/3, 2008, pp. 173–189.

[5]See Garth Le Pere and Francis Ikome, "Challenges and Prospects for Economic Development in Africa," Asia-Pacific Review, Vol. 16, Issue 2, Nov. 2009, pp. 89–114 and Greg Mills and Jonathan Oppenheimer, "Partners Not Beggars," *Time Europe*, Vol. 160, No. 2, 2002.

[6]Tsheola, "Global 'openness' and trade regionalism". and Richard Cornwell, "A New Partnership for Africa's Development?" *African Security Review*, Vol. 11, No. 1, 2002, pp. 91–96.

[7]Ngozi Okonjo-Iweala, "Africa's Growth and Resilience in a Volatile World," *Journal of International Affairs*, Vol. 62, Issue 2, Spring/Summer 2009, pp. 175–184.

age five—nearly 20 times the average 1 in 167 for developed regions. Life expectancy is 46.1 years compared to 77.7 years in the developed world. Only 14 doctors are available per 100,000 people compared to 290 in industrialized nations. More than 50 percent of Africans suffer from water-related diseases. Illiteracy among Africans is quite alarming. The rate of illiteracy is 40 percent for people beyond the age of 15.[8] Other notable facts further highlight the African situation. Of the forty most heavily indebted countries, thirty-three are in Africa. This is particularly disconcerting since it means that two-thirds of some fifty-five nations in Africa are mired in debt. Moreover, their per capita incomes are largely at the bottom of the world economic ladder; at present average per capita income is about $4300 (see Exhibit 3.2 below). No other continent comes close to such disturbing distinction. One observer reports "Africa is the great exception to the defining and otherwise global economic trend of recent decades: steady improvement in people's lives. In absolute terms, the head count of global poverty has fallen by 400 million. Some places (China, India) have seen steeper drops than others. Region by region, though, poverty has dropped everywhere, except in Africa. There, it continues to rise. Africa is making progress here and there, but far more slowly than are other poor regions. On virtually every measure, it is the outlier, the underperformer. Africa is the great development failure."[9] This impression prevails despite efforts to attenuate it, particularly the adoption and implementation of NEPAD.

The success of NEPAD is imperative, hence capacity innovation essential. "Africa is not simply an epicenter of economic failure; it is also the epicenter of a pervasive failure in what might be called human development."[10] The fate of some 900 million Africans is at stake. Successful implementation of NEPAD may be well served by elucidation of a conceptual framework for capacity innovation that offers a perspective on its complexity and its importance to development. It may be prudent to engage this perspective by starting at essentially square one and proceeding forward to a sort of capstoning scheme. What is capacity innovation's purpose, what's missing at this point in African economic evolution, what is the endgame that serves to galvanize events as they unfold and what might be done to put events on a winning trajectory?

---

[8]WHO, 2011.

[9]Clive Crook, "Wealth of Nations—When Economic Development Just Isn't Enough," *National Journal*, Vol. 34, No. 33–35, pp. 24–36.

[10]Nicholas Eberstadt, "The Global Poverty Paradox," *Commentary*, Vol. 130, Issue 3, Oct. 2010, pp. 16–23.

3.2 Capacity Innovation's Purpose: What It Needs to Accomplish

A statement of capacity innovation's purpose may be approached by way of historical thinking on the topic particularly as it relates to the development process itself. Various notions, theories if you will, of economic development attempt to characterize a systematic process of causal relationships among economic variables – Walter W. Rostow's stages of economic growth, David Ricardo's classical theory of economic growth, Karl Marx's historical materialism, Paul N. Rosenstein-Rodan's big push thesis, Albert O. Hirschman's strategy of unbalance, Paul A. Baran's neo-marxist thesis, Celso Furtado's dependency theory, and Andre Gunder Frank's dependency approach; all suffer from incapacity to fully and accurately represent and characterize across-the-board development among the world's assortment of countries.[11] For purposes of this dialogue Rostow's work offers interesting context – notwithstanding stage theory criticism and entanglements concerning continuity/discontinuity, continuity/periodization and stage discerning/periodization issues.[12]

Rostow postulates that investment in capacity innovation prompts the emergence of "leading sectors," the growth of which is thought to be instrumental in propelling the economy forward. The central thesis in Rostow's argument is that the growth of leading sectors can foster economic take-off, a process in which the scale of economic activity reaches a critical level and produces progressive and continuous positive structural transformation. In so postulating, Rostow has advanced a perspective on capacity innovation characterizing stages of economic growth, as demonstrated in Exhibit 3.1, in which all societies exist or pass through: the traditional society, emergence of the preconditions for take-off, the take-off, the drive to maturity, and the age of high mass-consumption.[13] Despite much criticism of the stage theory approach, it serves to provide important framing to the vexing problem of development.

---

[11] E. Wayne Nafziger, *The Economics of Developing Countries*, Second Edition, Englewood Cliffs, NJ: Prentice-Hall, Inc., 1990, pp. 76–96. [An insightful critique of these theories of economic development may be found here as well.]

[12] Nafziger, ibid. For readers unfamiliar with stage theory criticism, Itagaki's paper on the matter may be enlightening. See Yoichi Itagaki, "Criticism of Rostow's Stage Approach: The Concepts of Stage, System and Type," *The Developing Economies*, Volume 1, Issue 1, March 1963, pages 1–17.

[13] W.W. Rostow, *The Stages of Economic Growth: A Non-Communist Manifesto*, (Chapter 2, "The Five Stages of Growth—A Summary," pp. 4–16), Cambridge: Cambridge University Press, 1960, (adapted with permission of Cambridge University Press).

28 Towards an Analytical Policy Framework for Assessing Capacity Innovation

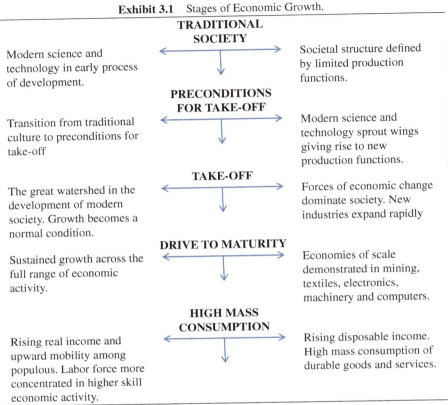

Exhibit 3.1 Stages of Economic Growth.

Source: W. W. Rostow, *The Stages of Economic Growth: A Non-Communist Manifesto*, (Chapter 2, "The Five Stages of Growth—A Summary," pp. 4–16), Cambridge: Cambridge University Press, 1960, (adapted with permission of Cambridge University Press). See also Gerald M. Meier, *Leading Issues in Economic Development*, New York: Oxford University Press, 1995, (adapted with permission of Oxford University Press).

The single most important stage for African countries arguably may be take-off as the road to this stage of development is a steady incline with little or no leveling until it is reached, marked by rich capacity innovation and perceptibly manifesting self-propelling energy thrusting the economy into and through maturity and on to mass-consumption. But since take-off requires systemic preparation, advent of the traditional society and emergence of the preconditions for take-off must materialize so as to give rise to take-off itself. Predicatory to economic preparation and beyond are strategic investment and growth of particular sectors in the economy so as to prompt an appreciable rise in the rate of productive investment.[14] The salient features of each stage are noteworthy. Appropriately, we begin with the traditional society. Conceptually, traditional society is rather fluid; economic, social and even political change abound, albeit, within limited production functions. Characteristic of a sort of pre-Newtonian view of the world, traditional society emerges subject to limited perceptions of laws of nature, and a systemic capacity for productive manipulation. Because of limitations on productivity, traditional society devotes a large share of its resources to agriculture. In this context, ad hoc technical innovations produce infrastructure improvements and a diffusion of new crops leading to increases in output.

Of course the veracity of such innovations is constrained by scientific achievement of the time. Attendantly, limited capacity innovation and manufacturing develop precipitating growth of industry and trade. The agricultural system prompts emergence of a hierarchical social structure, however with limited vertical mobility. Social organization is predicated upon a notion of intergenerational sameness, that is, opportunities available to offspring are largely consistent with those available to foreparent. This by no means prevents one from striving to improve one's lot during one's lifetime but clearly social structure limitations retard the speed of achievement.

Traditional society exhibits political manifestations as well, though characteristic of central rule. That is, political power in traditional society tends to portray a regional makeup emanating from those who own the land. From a historical perspective, landowners were the powerbrokers. This was clearly the case in the 'traditional society' construct, evidenced throughout the China dynasties, Middle East and Mediterranean civilization, medieval Europe and ancient Africa. Rostow cast these societies in a single grouping, if only to accommodate easy transition to elaboration of post-traditional societies, in which major characteristics of the traditional society, political, social and economic, change in such ways as to overcome particularly feudalistic practices[15] thus permitting normal forces of growth to take hold.

The second stage of growth engages societies as they emerge from the traditional model and begin development of the preconditions for take-off. In so doing, traditional society develops proficiency in exploiting the benefits of modern science, becomes efficient in fending off diminishing returns and

---

[14]"A rise in the rate of productive investment from, say, 5% or less to over 10% of national income (or net national product)." See Gerald M. Meier, *Leading Issues in Economic Development*, New York: Oxford University Press, 1995, p. 69.

[15]Jack Goody, "Feudalism in Africa?," *Journal of African History*, Vol. IV, No. 1, 1963.

embraces the rewards of the time value of money. Modern science and technology sprout wings, giving rise to new production functions: benefits accrue to both agriculture and industry; capacity innovation takes root, international competition expands; world markets flourish. As the pace of change among traditional societies varies, early developers hasten the maturation of late bloomers. Transition from traditional culture to the preconditions for take-off embraces in a very formal manner the notion that economic progress is not simply desirable but necessary to beget growth in the general welfare, increase educational achievement, encourage risk taking and entrepreneurship, promote institutional development, create comparative advantage—all giving rise to advances in modernization in transport, communications and manufacturing. However, vestiges of traditional society constrain the pace of this activity.

In many instances, traditional society persists alongside modern economic activity—in some cases a result of the natural order of modernization, in other cases motivated by political agenda of colonial or quasi-colonial power, or both. Although transition from traditional society to take-off involves significant changes in economic and social standards, a decisive and influential feature in this transition is often of a political nature. In this regard, amassing coalitions to build efficient and effective centralized government (the institutional mapping matter presented in Chapter 2) in opposition to the traditional regional interests of powerful landowners and/or the colonial power is a decisive feature of the preconditions period and is, essentially universally, a necessary condition for take-off. Put another way, sustained capacity innovation (take-off) will require a capitalist structure (new production functions), in which an efficient market mechanism is at work, and a definitive institutional mapping, in which efficient government institutions are in play.

The preconditions period is followed by what may be the single most important stage of growth and certainly the great watershed in the emergence of modern society—the take-off. At this critical juncture in the development of modernity, barriers and resistance to steady growth are finally overcome. Capacity innovation intensifies, forces of economic change dominate society, growth is normalized, time value of money and compound interest settle into financial habits and institutional structure. Innovation and surges in technological advancement provide a lifeline of fuel for continuous stimulus to take-off, accompanied by political innovations to support economic modernization. During take-off, the rate of investment rises to support new industries, thereby raising the demand for labor, support services, agricultural products, other manufactured goods, and new capital formation as well.

3.2 Capacity Innovation's Purpose: What It Needs to Accomplish 31

These activities yield increased income resulting in higher rates of saving but also channeling of savings into modern sector development. Entrepreneurship expands into broader industrial development spawning exploitation of unused natural resources and allocation of resources to highest and best use.

Entrepreneurship takes agricultural development to new heights as well introducing advanced growth and commercialization techniques for crops, leading to revolutionary changes in agricultural output—an essential condition for successful take-off. In time, economic, social and political structure of the society is transformed in such a way that a steady rate of growth is sustained. Such achievement is particularly demonstrative of G8 countries, although not necessarily or fully in the same way.

Following take-off, the drive to maturity begins. An extensive period of sustained growth becomes pervasive across the society's entire gamut of economic activity. Steady investment results in advanced capacity innovation and enables economic output to outstrip population increase. Innovation continues at a rapid pace: new industries accelerate; old industries level off; participation in the international economy ramps up, exported goods expand; comparative advantage rises, fostering increased competitiveness; and many imported goods are replaced by home manufactures. During the maturity stage, capacity innovation and economies are demonstrated across the range of economic activity – skillfully illustrated by the ladder of comparative advantage in Exhibit 3.2 – from resource-intensive (timber, coal, iron) to unskilled labor-intensive (textiles) to skilled labor-intensive (electronics)

Exhibit 3.2 Ladder of Comparative Advantage.

Source: Gerald M. Meier (1995). *Leading Issues in Economic Development,* New York: Oxford University Press, p. 458, by permission of Oxford University Press.

to capital-intensive (machinery) and ultimately to R&D and knowledge-intensive (computers).[16] While the economy may have raw materials or other supply constraints, its drive to maturity is a matter of economic choice or, political or societal priority rather than technological or institutional necessity. There is the notion that something on the order of sixty years is required in transitioning from the beginning of take-off to maturity. But in reality, this is difficult to predict. Timeliness in achieving comparative advantage and economies of scale across the full range of economic activity may vary across societies and indeed may depend upon economic and social commitment as well as political will. But perhaps it can be stated with a measure of dogmatism that societal transformation requires a broad range of success in energizing economic wherewithal and an even broader range of success in sustaining it. Only time will tell.

In time, the society shifts towards high mass consumption of durable goods and services. This stage of development is marked by mature capacity innovation and supported by rising real income among a large portion of the population. Household budgets accommodate basic food, shelter and clothing expense without significant strain. The structure of the workforce becomes increasingly urban and workers are spread plentifully across the range of economic sectors: resource-intensive, unskilled labor-intensive, skilled labor-intensive, capital-intensive, and R&D and knowledge-intensive as well. There is a very significant allocation of labor among higher skilled activity as take-off and the drive to maturity prompt an increasingly capable and experienced labor force—the ultimate result is a society endowed with upwardly mobile workers and an economic and social structure strongly middle class in character. Innovations and increasing efficiency in economic output provide lower-cost durable goods and services with the ability of a wide range of the population to afford washers, dryers, refrigerators, automobiles, homes, leisure time activity, and even disposable income for investment in securities and other income generating exploits. Notably, in the post-maturity stage, society allocates increased resources to social welfare in an effort to share the fruits of economic success with those less able to realize the benefits of modernity.

In light of the stages of growth landscape, where is Africa positioned? A reply to this question may be aided by further enlisting the aid of the ladder of comparative advantage. The ladder's framework is based on the normal 'ladder and queue' structure of development and world trade. Progression up the

---

[16] Gerald M. Meier, *Leading Issues in Economic Development*, New York: Oxford University Press, 1995, p. 458, (adapted with permission of Oxford University Press).

ladder and movement through the queue signal increasing capacity innovation beginning with natural resource-intensive exports and ultimately developing a capacity for R&D and knowledge-intensive output. Natural resource exports constrain a country to "natural" comparative advantage, that is, engagement in resource-intensive and unskilled labor-intensive production. Such activity tends to provide very limited comparative advantage invariably leading to heavy dependence on imported goods and services. Primary dependence on imports points to a serious weakness in the ability to achieve "created" comparative advantage, that is, the substitution of machines for labor and ultimately moving to R&D and knowledge-intensive exploits.

The ladder and queue process offers a formidable challenge to African development. Moving from natural resource-intensive exports, Africa's primary strength, to simple manufactures and then to more complex and high tech manufactures require relative factor endowment, i.e., highly productive land, labor, capital and entrepreneurship, that is generally well beyond the current capacity of African countries. Two inadequacies are particularly problematic: labor is plentiful but relatively unskilled and capital is notably lacking in degree and sophistication. So then, what may be proposed about Africa's position on the stages of growth landscape? It is apparent that the continent on the whole generally shows signs of movement beyond traditional society and even in some respects through preconditions for take-off. But the next bump up – take-off itself and beyond – has been a relatively insurmountable hurdle. The continent simply has not been able to sustain the forces of economic change necessary to create a lifeline of fuel for continuous, consistent and significant upticks in modernity. Further, inability to generate economies across the range of economic activity in that the continent has yet to develop much in the way of "created" comparative advantage, a clear signal of limited capacity innovation, will likely hold Africa at the door of economic take-off for some time to come.

While generalization about African countries holds validity in large part, there is evidence here and there, albeit limited, of migration in some fashion through the stages of development and up the rungs of the ladder of comparative advantage. The problem is that the stages show very weak development completion and footprints on the rungs are faint and at times resemble a disappearing act—first emulating live pulse, then flickering as if to signal stalled progression, and again pulsating with hope and aggression. The missing ingredient is a sustained heartbeat. Development and certainly the ability to traverse the stages and climb the ladder with continuously powered forward motion necessitates a sustained heartbeat and strongly so at that. Simply showing evidence of stage development clearly does not alone beget sustained development; indelibly footprinting the rungs with continuous firepower is requisite. Stage progression and rung advancement beget development and signal intensifying capacity innovation; mastering these is the key to sustainment. As important, a well-integrated economy in which all rungs show vibrant and sustained activity is the more efficient and preferred mode of development as, despite the level of development, countries typically display a range of human capital in need of employment and skills development across the range of economic activity as depicted by the ladder of comparative advantage. For the more advanced countries in particular, weakness or regression on any rung may prove especially problematic. The "post-subprime" U.S. economy exemplifies this condition. Well understood is that the subprime debacle cast the American economy into recession; not so well understood is that a hole in the country's ladder precipitously prolongs the recession making it exceedingly difficult for market forces to resurrect the economic vitality commonly observed. America has lost, or some feel given up, its comparative advantage in intermediate manufactures—skill levels and rungs on the ladder that must be maintained if the ladder is to remain intact. Importantly, capacity innovation linkage among the rungs ensures economic continuity among sectors particularly important to jobs sustainability and upward mobility of the labor force.

Despite the plausible economic argument for the flow of capital to low wage countries (for example, divestment in intermediate manufacturing in the U.S. and investment in this activity in, say, China) as reason for the hole, it is clear that when footprints on the rungs are faint or worse disappear, trouble is not just on the horizon but comes home to "roost." America has good success in correcting such misfortune in the past but its "post-subprime" struggles tell a different story. Pulling out of the downturn comes, but at what price. Sluggish recovery is symptomatic of an economy whose ladder is in distress; the hole created by missing rungs that once represented intermediate manufactures and employed an abundance of workers with skills to command middle income wages puts significant drag on the economy as market forces encounter tremendous impediments to redirecting capital flows to these rungs of the ladder. Workers once gainfully employed on these rungs become virtual spectators and in many instances take jobs at lower pay where available. The lesson to be learned here is that there can be no letup. Accomplishing stage and ladder preeminence is only part of the growth equation, sustaining this achievement is absolutely essential to avoiding economic turmoil. America's experience offers a tremendous window through which Africa may assess its plight. The continent not only has to achieve substantial growth but sustain

3.2 Capacity Innovation's Purpose: What It Needs to Accomplish

it as well. The stages of growth and the ladder of comparative advantage frameworks enlighten us about what needs to be done. For Africa, how to engineer lifting power and forward thrust to promote forces of change is the task ahead.

But before delving further into this matter, a necessary caveat is important. It is recognized that positioning the entire continent, some fifty-five countries, within the five-stage schema perhaps suggests that all of the economies tend to matriculate through the same series of stages or the same series of rungs. It is more likely that no single sequence either determines the development of all African countries or maps their individual progress over time. Clearly, to suggest that every African economy tracts the same path of development from beginning and ongoing is to ignore the complexities of development forces, cultural differences and the like.[17] But would it not be rather difficult to reason that a country may accomplish capacity innovation sufficient for takeoff, for example, without initiating any preconditions for such a maneuver? Wouldn't this be tantamount to walking before crawling, rowing without a paddle, burning brush without fire, passing laws without legislation, etc., etc.? These matters notwithstanding, characterizing African countries as essentially developing in line with the stage/rung mapping is a generality that is not necessarily unwarranted; African countries generally have not shown a propensity to/for doing otherwise. Mind you, this is not a criticism but rather a critique. Any form of consistent and sustained development and capacity innovation in African countries would be a welcome achievement whether initiated at stage 3, rung 5 or otherwise. It is not uncommon in development schemes to observe traditional society alongside modern transformations. And, a thoroughly integrated economy may well be one that bridges the full range of rungs on the ladder with plentiful natural resources and gainful returns from exports, labor-intensive activity to accommodate unskilled human resources, intermediate manufactures to employ skilled labor, capital-intensive investment for advanced skills and "created" comparative advantage, and at the very high end of economic prowess—knowledge-intensive pursuits specializing in R&D.

So what then does all of this really mean for Africa in terms of its position of development and trajectory for capacity innovation. Debatably there are some missing ingredients that facilitate the whole of the development process; these stand as "core challenges" to Africa's forward movement and indeed impediments to development. Clearly Africa must identify a means of

---

[17]Ibid., p. 71.

36   Towards an Analytical Policy Framework for Assessing Capacity Innovation

overcoming its "core challenges" while at the same time moving forward with stage development and ladder ascension. Above all, what really matters is that forward-looking capacity innovation takes root that is sustained over time.

3.2.1 What's Missing – A Look at the Core Challenges

Orchestrating capacity innovation is onerous under the best of circumstances; the structure and composition of African economics is clearly not gifted with the best of circumstances. Africa must surmount many core challenges not the least of which involve, as conferred in Exhibit 3.3, shortcomings with regard to

Exhibit 3.3   Core Challenges.

Comparative advantage (exports whose comparative costs are lower at home than abroad and imports whose comparative costs are lower abroad than at home)

• • •

Mini-states (compilation of 55 nations—mostly small; lack sizeable domestic markets to accommodate enterprise of any efficient scale)

• • •

Critical mass (smallness often has the unique problem of critical mass—too few resources, too few people; too few skills; too few innovators; too little capital)

• • •

Price-takers rather than price-makers (rely primarily on export of natural resources and import of manufactured products)

• • •

Lack of economic complementarity (export natural resources but produce little finished goods for intra-African consumption)

• • •

Inadequate intra-Africa communications systems (Africa to Europe but not Africa to Africans intra-Africa telecommunications system)

• • •

Failure of collective self-reliance (export-substitution policy meant import-substitution rather than the traditional export-led policy favored by the World Bank)

• • •

Economic integration (removal of barriers to commodity trade and impediments to intra-Africa factor movements)

• • •

Lack of international currency (no mutually acceptable transaction currency)

• • •

Good governance (guiding principles with implementing procedures and rules of law to assure observance and practice)

Source: Robert S. Browne, "How Africa Can Prosper," in David N. Balaam and Michael Veseth, editors, *Readings in International Political Economy*, Upper Saddle River, New Jersey: Prentice-Hall, Inc., 1996, (adapted with permission of Duke University Press).

comparative advantage, mini-states, critical mass, price-taker status, economic complementarity, intercontinental communication, hard currency, collective self-reliance, economic integration as well as good governance. These core challenges stand at the very heart of Africa's capacity innovation dilemma. A closer look at these would be instructive.[18]

3.2.2 Comparative Advantage

Successful capacity innovation will depend heavily upon Africa's ability to develop comparative advantage – the capacity to produce a particular good or service at a lower cost per unit than can be achieved in producing another good or service. Applying this to the country level, when country A can produce a good or service more cheaply than country B, it is said that country A has a comparative advantage. When country B can produce a good or service more cheaply than country A, then country B has a comparative advantage. The important principle derived from this is that a country should specialize in the production of goods and services in which it has a comparative advantage because such behavior results in the most efficient production and optimum general economic welfare.[19]

Comparative advantage is predicated upon a country's relative factor endowments, that is, natural resources, labor and capital—the inputs necessary for production. The pickle here is that African countries are struggling to achieve comparative advantage beyond natural resources. They have a relatively abundant factor in labor which should attract significant foreign direct investment and provide cheap labor production opportunities in foreign trade, but the skill level is so low as to render it uncompetitive with cheap labor in China, for example, where production opportunities are bid away from African countries. FDI to China is comparatively far more abundant. Thus with comparative advantage only in natural resource production, climbing the ladder of comparative advantage is a difficult proposition as resource-intensive

---

[18]For further perspective, see Benjamin F. Bobo and Hermann Sintim-Aboagye, *Neo-Liberalism, Interventionism and the Developmental State: Implementing the New Partnership for Africa's Development*, Trenton, New Jersey: Africa World Press, 2012, pp. 206–223, (adapted with permission of Africa World Press).

[19]See Robert S. Browne, "How Africa Can Prosper," in David N. Balaam and Michael Veseth, editors, *Readings in International Political Economy*, Upper Saddle River, New Jersey: Prentice-Hall, Inc., 1996, p. 280, (adapted with permission of Duke University Press). Also, the general reader will find very concise statements on comparative advantage in Christine Ammer and Dean S. Ammer, *Dictionary of Business and Economics*, New York: The Free Press, 1977, pp. 8 and 168.

activity alone does not produce sufficient propulsion without the adjoining forces of labor-, capital-, and knowledge-intensive activities to convert the natural resources to more profitable end-user products. When successful, this translates to foreign trade which can be a driving force in development.

3.2.3 Mini-States and Critical Mass

The problematic nature of foreign trade links to other impediments to capacity innovation. Browne underscores the matter of mini-states and economic irrationalism.[20] Africa is for the most part a compilation of very small nations. The largest, excluding Nigeria, are comparatively small relative to the United States, Russia or Japan—the largest G8 countries. Domestic markets of mini-states are generally too small to accommodate enterprise of any efficient scale, thus making it exceedingly difficult to sustain development initiatives. Smallness often has the unique problem of critical mass—too few resources, too few people, too few skills, too few innovators, and too little capital. Hence there is no engine of growth to spawn economic independence, and certainly not comparative advantage. An economic unit created as a mini-state of government and too small to achieve critical mass or an efficient scale of production is an irrational undertaking. The critical mass problem is exacerbated by trade barriers. With limited avenues to international markets and limited capacity to compete with particularly the G8, African countries simply are unable to take advantage of potentially lucrative trade opportunities to any significant degree.

The mini-states issue is particularly apparent when comparing, for example, the population, gross domestic product (GDP) and gross domestic product per capita (GDP per capita) of African countries to G8 countries. Exhibit 3.4 offers illustrative statistics. Forty-five of the 55 African countries have populations below the smallest G8 country—an average of 9.4 million compared to Canada's approximately 35 million. No African country measures up GDP-wise—the largest, Nigeria, with about $450 billion compared to Canada's $1.45 trillion. In fact, excluding Nigeria, Egypt and South Africa, the total GDP of the remaining 52 African countries, approximately $1.8 trillion, is only comparable to the GDP of Italy, the seventh largest G8 country. And, GDP per capita comparison is perhaps as profound particularly when considering countries in Africa with populations above 30 million—the average GDP per capita for these countries standing at $4,298 compared to $35,825 for the G8.

---

[20]Browne, op cit., p. 280.

3.2 Capacity Innovation's Purpose: What It Needs to Accomplish 39

Exhibit 3.4 The Mini-states Perspective African Countries by Comparison to G8 Population, GDP and GDP per capita.

Country	*Population (million)	GDP ($ trillion)	GDP per capita ($)
G8			
United States	316.67	15.66	49,800
Russia	142.50	2.51	17,700
Japan	127.25	4.62	36,200
Germany	81.15	3.19	39,100
France	65.95	2.25	35,500
United Kingdom	63.40	2.32	36,700
Italy	61.50	1.84	30,100
Canada	34.57	1.45	41,500
Africa			
Nigeria	170.90	.45	2,700
Ethiopia	93.88	.10	1,200
Egypt	85.30	.54	6,600
Congo, Democratic Republic	75.51	.03	400
South Africa	48.60	.58	11,300
Tanzania	48.26	.07	1,700
Kenya	44.04	.08	1,800
Algeria	38.09	.27	7,500
Sudan	34.85	.08	2,400
Uganda	34.76	.05	1,400
Morocco	32.65	.17	5,300
Ghana	25.20	.08	3,300
Mozambique	24.10	.03	1,200
Madagascar	22.60	.02	1,000
Cote d'Ivoire	22.40	.04	1,700
Cameroon	20.55	.05	2,300
Angola	18.57	.13	6,200
Burkina Faso	17.81	.02	1,400
Niger	16.90	.01	900
Malawi	16.78	.02	900
Mali	15.97	.02	1,100
Zambia	14.22	.02	1,700
Senegal	13.30	.02	1,900
Zimbabwe	13.18	.007	500
Rwanda	12.01	.02	1,400
Chad	11.19	.02	2,000
Guinea	11.18	.01	1,100
South Sudan	11.09	.01	900
Burundi	10.89	.006	600
Tunisia	10.84	.10	9,700

(Continued)

Exhibit 3.4 Continued

Country	*Population (million)	GDP ($ trillion)	GDP per capita ($)
Somalia	10.25	.006	600
Benin	9.88	.02	1,700
Togo	7.15	.007	1,100
Eritrea	6.23	.004	800
Libya	6.00	.09	13,300
Sierra Leone	5.61	.008	1,400
Central African Republic	5.17	.004	800
Republic of the Congo	4.50	.02	4,700
Liberia	3.99	.003	700
Mauritania	3.44	.008	2,100
Namibia	2.18	.02	7,800
Botswana	2.13	.03	16,800
Lesotho	1.94	.004	2,000
The Gambia	1.88	.004	1,900
Guinea-Bissau	1.66	.002	1,100
Gabon	1.64	.03	17,300
Swaziland	1.40	.006	5,300
Djibouti	.79	.002	2,700
Comoros	.75	.0009	1,300
Equitorial Guinea	.70	.03	20,200
Western Sahara	.54	.0009	2,500
Cape Verde	.53	.002	4,100
São Tomé and Principe	.19	.0004	2,300
Seychelles	.09	.002	26,200
†Saint Helena Ascension, and Tristan da Cunha (UK)	.007	.0002	2,500

Source: CIA World Factbook 2012.
*Population figures are 2013 estimates; GDP and GDP per capita are 2012 purchasing power parity estimates.
†Excluded from total number of African countries as it is a UK territory.

These statistics reveal the daunting task ahead of NEPAD. If African countries are to establish themselves as significant economic entities within the world body of nations, the New Partnership for Africa's Development will have to raise GDP levels at least reasonably so to measure up. While setting their sights on G8 performance may be a bit too ambitious at the moment, it creates a sizeable margin for error. This notwithstanding, the G8-Africa comparison offers an instructive perspective on the mini-states/critical mass predicament confronting capacity innovation in African development.

3.2.4 Price-Takers

Magnifying the problem of critical mass is that African countries are in the difficult position of being price-takers rather than price-makers. Browne argues that this is the case in part because they rely primarily on export of natural resources and import of manufactured products.[21] Frankly, this is indicative of the normal 'ladder and queue' structure of development and world trade. Exhibit 3.2 above lends perspective here as it illustrates what can be expected in progressing up the ladder and moving through the queue. Natural resource exports constrain a country to "natural" comparative advantage, that is, engagement in resource-intensive and unskilled labor-intensive production; such activity tends to provide very limited comparative advantage. Primary dependence on imports point to a serious weakness in the ability to achieve "created" comparative advantage, that is, the substitution of machines for labor and ultimately moving to R&D and knowledge-intensive exploits—a process occasioned by attainment of capacity innovation.

G8 countries abound with "created" comparative advantage as they have achieved development at the upper rungs of the ladder. Price-maker status is a simple result of this level of development. African countries on the other hand being endowed largely with "natural" comparative advantage with limited capacity to produce manufactured products tend to have far less opportunity to control the price of goods and services; the result of their position at the lower rungs. This is quite apparent considering that "created" comparative advantage obtains production of more complex and high tech manufactures, e.g., upscale electronics, machinery, computers, etc., while "natural" comparative advantage relegates to the production of agriculture, timber, bauxite, oil, etc. Of course in the very unique case of oil, price control rests on the side of oil producers because of a cartel arrangement. It exemplifies perhaps the most effective use of such an arrangement. Though Nigeria, for example, is an oil producer, a member of the OPEC oil cartel and has some 170 million inhabitants, it has yet to develop much in the way of "created" comparative advantage. Even though labor there is plentiful, it is relatively unskilled and while natural resources (oil, natural gas, tin, iron ore, etc.) tend to be plentiful as well, their export alone will not lead to "created" comparative

---

[21] Ibid., pp. 280–281.

advantage. Nor will a large population independently suffice to obtain critical mass; Nigeria has too few skills and too few innovators—very serious impediments to capacity innovation. And, Nigeria has nearly a half trillion dollar economy. With a larger contingent of innovators and the support of more advanced skills, the capital markets would allocate resources to economic innovations that result in "created" comparative advantage which would lead to progression up the "ladder," advancement through the "stages," movement towards an "endgame," and the like. Such accomplishments engender price-maker capacity. Hence the job for NEPAD becomes more complex; it has to create pricing power not just in Nigeria where the task is perhaps less onerous but throughout the whole of Africa where "created" comparative advantage and critical mass are far more in doubt.

3.2.5 Economic Complementarity

A further obstacle to NEPAD is the lack of economic complementarity. The economies of African countries, particularly neighboring ones, do not complement each other very well. Most African countries produce mineral or agricultural commodities that are exported overseas. These products are rarely used in the unprocessed form and few African countries have the capacity to convert them into finished or consumer goods on a significant scale,[22] hence the critical need for capacity innovation. The demand for consumer goods provides very attractive trade opportunities among nations; trade between and among African countries is of particular concern here. The limited capacity to produce consumer goods impedes opportunities for comparative advantage and thereby thwarts intra-African trade. Unable to supply mutual needs or to counterbalance mutual deficiencies—capacities necessary for economic complementarity, African countries lose the power of development that these capacities facilitate.

Yang and Gupta note that high external trade barriers and low resource complementarity among African countries limit both intra- and extraregional trade. And small market size, high trading costs and inadequate transport facilities make it difficult to increase regional trade which in turn limits Africa's output. Africa produces only two percent of the world's output yet has 12 percent of the world's population.[23] Overcoming the economic

---

[22]Ibid., pp. 281–282.
[23]Yongzheng Yang and Sanjeev Gupta, *Regional Trade Arrangements in Africa: Past Performance and the Way Forward*, African Development Bank, Blackwell Publishing Ltd., 2007, pp. 399–431.

complementarity problem may well depend upon increased trade flows which link very directly to elevated output which in turn depends directly upon capacity innovation suited to Africa's needs. Clearly the challenges for African development are quite profound, but the scope of the challenges is not yet fully disclosed.

3.2.6 Communication

The inability of Africans to efficiently communicate among themselves given the availability of communications technology further hampers capacity innovation. Communications between African countries and the outside world are often less problematic than direct communications among African countries themselves. Absence of an African telecommunications superhighway, long within the capacity of communications technology, has prevented development of an intra-continental system of linkages that would more efficiently facilitate capacity innovation among African countries.[24] In the current information revolution, Africa lags behind other regions of the world. Of 1,270 million worldwide fixed mainline telephones in 2006, Africa's share was less than 2% while Asia stood at 48%. In Latin America, Brazil for example had a higher number of telephones than the whole of the African continent. In terms of fixed mainline penetration, Africa was 3.1 per 100 inhabitants compared to a world average of 19.5 per 100 inhabitants. Fixed lines are concentrated in just six countries on the continent – Algeria, Egypt, Morocco, Nigeria, South Africa and Tunisia—accounting for about 80 percent of all fixed mainline telephones in Africa. The only encouraging sign is the African mobile cellular market which has grown at about 49 percent annually during the last decade. Africa's mobile penetration of 22 per 100 inhabitants compares to 29 per 100 in Asia, 94 per 100 in Europe and 62 per 100 in the United States. Beyond fixed line and cellular communication, internet use has been quite limited as internet infrastructure to provide e-mail communication is too slowly developing for the masses.[25] Computer-based networking along with other forms of communication that would more efficiently unite Africans is important input to capacity innovation. Here again, NEPAD must overcome a core challenge. Improving access to communication technology for the

---

[24] Browne, "How Africa Can Prosper," p. 281.
[25] See Michael Enowbi Batuo, "The Role of Telecommunication Infrastructure in the Regional Economic Growth of Africa," http://mpra.ub-muenchen.de/12431/, MPRA Paper No. 12431, posted 30. December 2008/18:47, 22. June 2008, pp. 7–9.

African masses is requisite if it is to play the supportive role in capacity innovation required of it.

3.2.7 Collective Self-Reliance – Economic Integration

Browne also underscores the notion of collective self-reliance. Alarmed by their inability to reverse the persistently declining economics of the continent, in the late 1970s-early 1980s African countries adopted a sort of export-substitution strategy. As such, Africa was to pursue more direct production of goods it needed rather than producing goods for export and using the proceeds of the sales to purchase imports. This was seen as giving Africa a sense of self-reliance and hopefully a clearer path to capacity innovation. Further, recognizing the critical mass problem African governments were called upon to create regional markets in a move towards regional economic integration, thus making capacity innovation a matter of collective self-reliance. The Economic Community of West African States (ECOWAS) and the Preferential Trade Area of Eastern and Southern Africa (PTA) are essentially prototypes of this initiative but have had little success.[26]

The collective self-reliance initiative raised the ire of the donor community and particularly the World Bank. As the export-substitution strategy was effectively an import-substitution program, Browne contends that the World Bank rejected it because of its departure from the Bank's export-led development philosophy. Further, according to Browne, the World Bank in concert with the wishes of developed countries "find it in their interests to keep the African countries as mainly exporters of raw materials and importers of manufactured items rather than to assist them in becoming tenacious industrial competitors, as the newly industrialized countries of East Asia have become."[27]

The importance of regional economic integration should not be minimized in the overall scheme of encouraging activities that foster capacity innovation. It is not an issue that can take a back seat to other development priorities, particularly in the African situation, with high potential for end-game success. If outside interests are to prevail in controlling African capacity innovation strategy, economic integration is perhaps their strongest medium of detainment. Collective will is an all-powerful force of resistance; a force that is largely unknown to the continent. Should African economies achieve integration, success of the continent's capacity innovation strategy and ultimately

---

[26] Browne, "How Africa Can Prosper," pp. 283 and 286.
[27] Ibid., pp. 283.

3.2 Capacity Innovation's Purpose: What It Needs to Accomplish 45

the move to tenacious industrial competitor status may be all but assured. To be sure, this may be an enormous challenge for NEPAD, perhaps its most engaging test. Mattli makes this poignantly clear in arguing that successful integration requires two conditions to be met – a need to encourage changes in the extent and structure of markets as well as a desire to encourage such changes. Put another way, there must be a need to integrate economies as well as a desire to integrate economies. These are effectively demand side and supply side conditions. On the one hand, they involve regional institution-building in an effort to internalize externalities and costs associated therewith as a result of trade and investment within a group of countries. Externalities arise from uncertainty and risk faced by market players when interacting with foreign firms and governments, and subjugate to regional rules, regulations and policies which alleviate the costs of these externalities. On the other hand, they speak to a willingness by political leaders to accommodate demands for economic integration. Leaders value economic autonomy and may be unwilling to sacrifice national sovereignty if the expected marginal benefit of integration is perceived as minimal. Hence the desire to integrate simply may not arise. Even in times of clear economic difficulty—as in the African case—leaders may be unwilling to indulge economic integration simply out of fear of losing political control. This may be NEPAD's Achilles' heel. If economic integration is necessary for successful capacity innovation, is this a core challenge that defies resolve?[28]

3.2.8 Currency

Contributing to the challenges to African economic development, Browne contends, is the lack of an international currency. This is a rather serious problem among Third World countries on the whole and certainly is an impediment to intra-African trade. This is no less a problem for trade between Africa and the outside world. With no mutually acceptable transaction currency and the non-convertibility of most African currencies, intra-African trade as well as international trade must be conducted in hard currency. Because Africa has limited means to generate hard currency, that is, to produce goods and services that can be sold for dollars, pounds or euros, not only is trade impeded but imports of commodities that support economic development activities are restricted as well.[29] This places African countries in a veritable catch-22. Their

---

[28] Walter Mattli, "Explaining Regional Integration Outcomes," *Journal of European Public Policy*, 6:1, March 1999, p. 3.
[29] Browne, "How Africa Can Prosper," p. 282.

development is constrained because they lack sufficient hard currency; they lack sufficient hard currency because their development is constrained.

It has been suggested that a dual currency system would best serve Africa's needs—an intra-African currency and a foreign exchange reserve fund. These would ameliorate to some extent Africa's catch-22 predicament. The notion is that, in the case of the intra-African currency, there would be a single African currency circulating only within the confines of the continent to facilitate the exchange of goods and services among African countries. This would also serve well as an apparatus to assist regional integration and further to facilitate capacity innovation. Of course intra-African issues related to the movement of the factors of production, monetary and fiscal policies, as well as exchange rates and currencies convertibility among African countries in establishing the intra-African currency would have to be worked out. The currency would have a par value with internationally negotiable currencies and convertible thereto. Administration of the intra-African currency would require establishment of a supranational authority, presumably a NEPAD subunit or an African Central Bank. Concomitantly, an intra-continental foreign exchange reserve fund would be arranged to facilitate international import/export activities and regional economic integration pursuant to intra-national needs. Monies originating outside the continent would flow into this fund to support foreign business transactions and importantly to provide the power of collective foreign exchange earnings in accommodating the import/export needs of the continent. Here again, an administrative body charged with this specific undertaking is requisite.[30]

A third means of addressing financial matters would serve Africa's needs as well. NEPAD will incur a good deal of administrative costs. An indigenous investment fund enabled and supported by indigenous corporations would facilitate the ongoing administration of NEPAD and assist in avoiding lapses in organizational support by African countries that have been problematic in past collective development efforts. This fund would also backstop the foreign exchange reserve fund. It would provide a secondary means of supporting import/export activities and development projects. In supporting this effort, the indigenous corporations may be particularly concerned about the potential free-rider problem: firms may not be as receptive as expected since support of the fund is essentially a fee; an individual corporation receives only 1/nth of the benefit generated by its support. The free-rider problem would be solved

---

[30] Mammo Muchie, "Wanted: African Single Currency," *New African*, Issue 407, May 2002.

3.2 Capacity Innovation's Purpose: What It Needs to Accomplish 47

by mandatory participation with a prescribed nominal fee. This would directly abate the free-rider problem as firms would be required to participate in order to gain access to the foreign exchange reserve fund. The benefits of the fund and the desire to maintain competitiveness, i.e., to receive foreign exchange for imports crucial to supporting production activities,[31] are strong motivators for participation.

3.2.9 Governance

Governance is clearly problematic on the continent. The importance of the governance factor cannot be overstated. It is quite literally the glue that binds the components of development and certainly will dictate the process of capacity innovation. Without a radical change in governance throughout Africa, no amount of capacity innovation initiatives will have a significant chance of succeeding. Inefficient institutions, rule of law violations, human rights abuses, restricted freedom and the like point to serious flaws in Africa's governance mechanism. Thabo Mbeki underscores the matter quite ably in noting that "good governance on our continent, comprehensively understood, is of fundamental interest to the peoples of Africa."[32] Africa must ensure "that measures for good governance are put in place through which our governments are accountable to their peoples; that best practices are agreed upon and put in place for economic and political governance."[33] Amartya Sen, the Nobel laureate, offers a conjunctive comment, "development requires—and, indeed, is inseparable from—greater freedom."[34]

Hilton L. Root in his book *Small Countries, Big Lessons: Governance and the Rise of East Asia* confides that discussions on good governance arouse the sharpest disagreements and inspire the greatest introspection. But socioeconomic welfare depends upon it and progress toward African development clearly hinges on progress in good governance.[35] Logical and rational thought,

---

[31] Benjamin F. Bobo, *Rich Country, Poor Country: The Multinational as Change Agent*, Praeger Publishers, Westport, CT, 2005, pp. 175–176.

[32] Thabo Mbeki, "Mbeki: African Union is the Mother, NEPAD is Her Baby," *New African*, Issue 415, February 2003, p. 44.

[33] Thabo Mbeki, "New Partnerships for Africa's Development," *Presidents & Prime Ministers*, Vol. 10, Issue 6, November/December 2001, pp. 30–32.

[34] Philippe Legrain, "Africa's Challenge," *World Link*, Vol. 15, Issue 3, May/June 2002, p. 3.

[35] Hilton L. Root, *Small Countries, Big Lessons: Governance and the Rise of East Asia* (see Chapter 10, "The Search for Good Governance"), New York: *Oxford University Press*, 1996, p. 145.

based upon experiential observations, judgment, wisdom, and the like, would seem to support the notion that good governance largely evolves from an institutional mapping that prescribes as well as implements an organization of governmental decision-making units complete with a government philosophy or set of guiding principles (a constitution), procedures for implementing the guiding principles, and rules of law to assure that the guiding principles and procedures are observed and practiced. The degree to which this mapping affords an efficient and effective outcome depends largely upon the nature of the principles, procedures, and rules of law, and how these elements are operationalized. Importantly, the key task here surrounds the ability of African governments to deliver what they promise and to provide their citizenry opportunities to realize their expectations.[36]

To be sure, good governance derives from not one but several interdependent or mutually supporting elements of government structure. Good rules/constitution, good policy apparatus, good institutions, good political feasibility, and good government capacity all work in concert to ensure an accountable, predictable, and transparent policy environment. Absent these elements, achieving successful capacity innovation outcomes will be in a practical sense improbable. This is particularly relevant in the African case. Flawed processes occasioned by nontransparent and discretionary decision making, weak institutional capacity particularly in the areas of regulation, service delivery and social spending, and the incapacity to achieve the balance between private and social costs and benefits operate to impede economic activity. Critical to stimulating robust economics is ensuring that the regulatory environment is supportive of economic activity.[37] This has been a daunting challenge for Africa as some or all of the elements of good governance are in distress, and moreover, an effective and efficient interplay of the elements has not been achieved. NEPAD must ensure that African governments are sufficiently institutionally structured to enable them to efficiently and effectively support capacity innovation. This is indeed a daunting task but absolutely necessary to ensure that the glue that binds the components of capacity innovation has the strength to do the job.[38]

---

[36]Ibid., p. 147.

[37]Trevor A. Manuel, "Africa and the Washington Consensus: Finding the Right Path," *Finance & Development* (The International Monetary Fund), Vol. 40, No. 3, September 2003, pp. 19–20.

[38]For further reflection on the governance matter, see Benjamin F. Bobo and Hermann Sintim-Aboagye, *Neo-Liberalism, Interventionism and the Developmental State*, pp. 220–223.

3.2.10 The Capacity Innovation Endgame

In engaging the core challenges, NEPAD endeavors to foster the capacity innovation process in the guise of the First World model. As it decrees, "people-centered development and market-oriented economies are on the increase," and NEPAD "is about consolidating and accelerating these gains."[39] The First World capacity innovation process has very much a humanistic character—people acting in individual decision making giving rise to a system of collective voices expressing preferences for goods and services and providing employment of capital and labor. These features characterize a marketplace where they interact as an "invisible hand" to guide the workings of the market. The undergirding of the First World model is a neo-liberalism philosophy as denoted in Exhibit 3.5 and a profound accomplishment of capacity innovation, evident across the entirety of the ladder of comparative advantage from resource-intensive to R&D and knowledge-intensive economic activity. Inherent in the model is a clear endgame—satisfying human wants—the pursuit of which has produced vibrant economic development. As with the First World model, under the NEPAD approach to development the endgame is a critical feature of the development equation. Exhibit 3.5 outlines an endgame perspective in the African context, expropriating parallels to First World characteristics, which attaches a personal character to the capacity innovation framework in asserting an underlying mindset or "invisible hand" at work through assisted capitalism and assisted commerce. All development processes exhibit a philosophical leaning endeared with an intended or expected outcome. An efficient outcome will require NEPAD, a form of government intervention assuming a significant state role in development with powers to assist market forces and through such apparatus assist commerce itself, to fully recognize the benefit of market directed development particularly since it aspires to promulgate a continent of market-oriented economies.

NEPAD may be well advised to be mindful of prevailing thought on interventionist policy. Two perspectives are particularly instructive. Krugman supports the notion that interventionist policies might lead to more optimal results than those produced by markets but feels that politics are as imperfect as markets. He warns that pursuing strategic policies could be counterproductive and result in a misallocation of resources; thereby encouraging inefficient investment. He further cautions that predicting outcomes, as would be required to make practical sense of interventionist policies, would be problematic in

---

[39] See the New Partnership for Africa's Development (NEPAD), October 2001.

Exhibit 3.5 The Endgame Perspective Philosophy to Outcome.

Underlying Philosophies / Schools of Thought
- First World/Neo-Liberalism/Market Forces
- NEPAD/Interventionism/Developmental State

↓   ↓   ↓

Socio-Economic Constructs
- Capitalism/Laissez-faire/Free Enterprise
- Assisted Capitalism/Assisted Commerce

↓   ↓   ↓

Operating Frameworks
- Market Resource Allocation/Invisible Hand
- Assited Resource Allocation/NEPAD

↓   ↓   ↓

Capacity Development
- Technology Transfer/Skills Upgrade
- Economic Innovation/Climbing the Ladder

↓   ↓   ↓

Endgame
- Sustained Capacity
- Sustained Development
- Sustained Provision for Human Needs
- Sustained Satisfaction of Human Wants

complex strategic environments as prevalent in many industries. This matter would be even more profound in the context of nations as NEPAD engulfs fifty-five such environments. Stiglitz is concerned whether development economics underestimates the role of markets and differs with the notion that the problems of development may be ascribed primarily to market failures. He does not feel that the primary recipients of the developmental models should be governments who replace absent and/or imperfect markets and guide the economy towards a more efficient allocation of resources. Alternatively a

3.2 Capacity Innovation's Purpose: What It Needs to Accomplish

broader scope should be embraced, as he imagined, with more actors engaged in developmental efforts.[40]

To be sure, "Merging developmental state ideology and interventionist strategy with market forces arguably presents a dynamic that NEPAD must necessarily navigate while at the same time accommodate. History shows that developmental states have been staged upon active markets and will likely be the case going forward. Interventionist strategies whether undertaken in light of market failure or simply as corrective action to improve market performance," e.g., America's $700 billion dollar bailout of banks and auto manufacturers, "are from time to time necessary."[41] "The key for NEPAD is to avoid the trappings of unsustainable schemes and structures whether they result from misdirected developmentalism, overpowering interventionism, or an inefficient market mechanism."[42]

Spawning market-oriented African economies, as NEPAD desires, will necessarily derive from capacity innovation stemming from development philosophy embedded in the neo-liberalism school of thought: the give-and-take of the market produces the most efficient allocation of resources and provides maximum growth and development. A laissez-faire mindset underpins the emergence of a free enterprise system in which allocation of resources to highest and best use stimulates technology transfer and skills development necessary for capacity innovation and economic enrichment. State facilitation of a free enterprise system through assisted resource allocation must encourage activities such as foreign direct investment, capital diffusion, new capital formation, infrastructure development and the like to promote economic innovation and movement up the ladder of comparative advantage. Such resource mapping is requisite to enabling capacity innovation particularly considering the African situation. The entirety of philosophical perspective, socio-economic construct, operating framework and capacity development as informed by Exhibit 3.5 leads to a definitive endgame: sustained satisfaction of

---

[40] See Ivan Lesay, "How 'Post' is the Post-Washington Consensus?" *Journal of Third World Studies*, XXIX, 2, 2012. p. 188 and p. 184.

[41] See Karl Botchway and Jamee Moudud, "Neo-Liberalism and the Developmental State: Consideration for the New Partnership for Africa's Development," in Benjamin F. Bobo and Hermann Sintim-Aboagye, Eds., *Neo-Liberalism, Interventionism and the Developmental State: Implementing the New Partnership for Africa's Development*, Trenton, NJ: Africa World Press, 2012, pp. 40–41.

[42] Ibid., 41.

human wants; a result subsuming sustained capacity, sustained development and sustained provision of human needs.

Of course achieving a sort of economic capacity innovation parallel to the First World model as NEPAD essentially aspires, requires capacity innovation strategy capable of orchestrating such a feat. NEPAD imagines that as a super-organization with the backing of fifty-five nations, it has the standalone wherewithal to get the job done. Left to its own devices, what might NEPAD do to create First World-like capacity innovation? This is the ultimate question.

3.2.11 What Might Be Done: The Path to Capacity Innovation

From a standalone perspective, NEPAD acting with the collective power of the nations of the continent and charged with orchestration, initiation and oversight of future development is effectively entrusted with charting a path to capacity innovation. Exhibit 3.6 charts such a path quite literally beginning with the task of capital diffusion throughout the continent. Capital diffusion principally involves new capital formation – the basis of future production. It is useful to recognize that capital diffusion is far more complex than it may appear on the surface as it requires the full range of economic growth support elements including good governance – political, social and institutional framework – and reasonably plentiful factors of production—raw materials, skilled labor (cheap labor may be a substitute as a starter), capital, entrepreneurship, and foreign exchange. With this in mind, NEPAD acts as principal facilitator of capital diffusion through encouragement of, and directing where practical, government feasibility and investment in high net present value projects. NEPAD develops and implements uniform operating policies and practices, and provides oversight and monitoring of capital diffusion activities to ensure observance of operating dictum.

As the path emerges, African entities (especially joint ventures where asset pooling is advantageous and more efficient) initiate production projects in sectors targeted by NEPAD for development with a view towards creating comparative advantage. NEPAD must exercise caution here as targeting has potential downsides invariably owing to the usual application of one or more forms of industry protection: elimination or restriction of competition, provision of subsidies, taxpayer support, support of insufficient market size, support of undeveloped local industry, insensitivity to harm to downstream industry, and the like. Protecting an industry may eliminate the competition it must face in the short run, but long-run advantages such as development

3.2 Capacity Innovation's Purpose: What It Needs to Accomplish

Exhibit 3.6 Path to Capacity innovation Model of NEPAD Facilitation A Standalone Perspective.

Facilitate capital diffusion through encouragement of private and public new capital formation throughout continent, and development and implementation of uniform policies and practices, and provision of oversight and monitoring of capital diffusion activities.

Support development of comparative advantage through encouragement of economic sector targeting and development of uniform policies and practices to accommodate wholly owned as well as joint venture projects.

Encouragement of collective export-substitution strategy founded upon collective self-reliance and creation of regional markets and trade flows to promote and support regional integration.

Promote economic complementarity through development of uniform policies and practices to encourage production of goods and services to supply the mutual needs of African countries.

Promote development of price-maker status through development and implementation of collective policies and practices that encourage and support intra-continental competition, market-oriented activities, technological progress and gains in productivity.

Promote development of an African telecommunications superhighway through development of policies and practices that encourage such investment.

Develop infrastructure plan and encourage support of the development of infrastructure to accommodate creation of comparative advantage, regional economic integration, economic complementarity and intra-continental communication.

Develop policies and practices to encourage support of a special intra-continental investment fund.

Facilitate abatement of mini-states and critical mass problem through development of policies and practices that encourage investment activities that raise GDP levels.

Encourage intra-continental governance best practices to ensure operational uniformity, institutional validity, feasibility and organizational integrity.

Organize intra-continental security force to ensure observance of rule of law.

Source: Benjamin F. Bobo and Hermann Sintim-Aboagye, *Neo-Liberalism, Interventionism and the Developmental State: Implementing the New Partnership for Africa's Development*, Trenton, New Jersey: Africa World Press, 2012, pp. 206–223, (adapted with permission of Africa World Press).

of comparative advantage may not materialize as exposure to a competitive environment appears to be necessary to develop internationally competitive firms.[43]

---

[43] See Benjamin F. Bobo and Hermann Sintim-Aboagye, Eds., *Neo-Liberalism, Interventionism and the Developmental State: Implementing the New Partnership for Africa's Development*, Trenton, NJ: Africa World Press, 2012, pp. 234–236.

Continuing along the unfolding path, NEPAD promotes collective export-substitution production activities with particular attention to fostering regional integration in an effort to address Africa's critical mass problem. Intra-continental investors respond by organizing production facilities and supply of goods and services to support regional integration. These efforts lead to development of economic complementarity as NEPAD puts forth policies that encourage production of goods and services to supply the mutual needs of African countries. As capital diffusion strategy gives rise to comparative advantage, regional integration and economic complementarity, forces are set in place that spawn price-maker capacity within the collective facility of African countries. This is further enabled by encouragement and support of intra-continental competition, market-oriented activities, technological progress and gains in productivity.

Continued attraction of investment and sustained benefits of capital diffusion throughout the continent necessarily require efficient communication among the units of production and avenues of trade flows. NEPAD acts to stimulate investment to promote development of a telecommunications superhighway. Investors so inclined will respond to the call through initiation of communication projects that specifically link African countries throughout the continent via a network of communication technologies. Realization of the entire package of benefits orchestrated by NEPAD will be predicated upon development of a range of infrastructure, telecommunications included, to support the efficient diffusion of capital across the continent. Intra-continental roadways, power grids, railways, seaports, water supply and the like will command resources that NEPAD as an orchestrating apparatus can more efficiently obtain through encouragement of collective African government investment. Infrastructure development will especially benefit from the mutual cooperation of African governments as infrastructure projects may be exceedingly costly with any single user receiving only 1/nth of the benefit generated by the project. It would not be cost effective for a single African country to undertake intra-continentally oriented projects. However, a special intra-continental investment fund supported by the full cadre of African governments and even African firms would provide necessary capital to support infrastructure projects.

In total, the capital diffusion effort holds very serious promise for abating the mini-states and critical mass conundrum and raising GDP levels. Rising GDP over time means an exit from failed development initiatives and movement towards sustained development and economic innovation. To secure sustained development NEPAD encourages, and directs where practical, implementation of governance best practices among African governments. Efficient development requires the support of efficient institutions to ensure operational uniformity. Best practices lead to institutional validity, feasibility and integrity. As a final task, NEPAD organizes an intra-continental security force to assure observance of rule of law. Invariably the need for such becomes a reality. Moreover, investor confidence will be heightened by such assurance.

The path to capacity innovation therefore has many and varied hurdles. Traversing the hurdles will require a definitive approach to building necessary capacity to initiate development and sustain it through take-off and beyond. Exhibit 3.6 outlines such an approach from the perspective of situating a super-organization in a facilitating role enlisted and empowered by intra-continental consensus of national governments. In this role, the facilitator in a standalone mode engages the tasks as prescribed with a view towards promoting outcomes, of course being mindful of potential downsides of interventionist policy.

3.2.12 Closing Comments

Having said all that's been said, a sort of verse of caution may be in order. David Landes situates thinking about capacity innovation and indeed development success in a very instructive manner. Capacity innovation takes time, hence, are developing countries prepared to wait many decades and perhaps even a hundred years to reach sustained capacity—a necessary condition for developed status? If put directly to countries so defined, such a question would likely evoke outrage. Why wait for what seems to be an eternity when others have achieved such status in a much shorter period? Perhaps by way of example a simple note, however disquieting, may lessen the crassness of the question: Japan endured such a development period. The capacity innovation process is a time-consuming interaction of people, places and things; an interaction that occasions events; events that appear as stages, rungs, challenges, endgame and a path to the ultimate objective—sustained capacity innovation. Of course there are many alleged reasons why events have led to success or failure. Successful countries are said to be early comers to the game, subscribers to market allocation of resources, beneficiaries of staples theory (vent for surplus), believers in the notion of comparative advantage, advocates of private ownership and the accumulation of wealth, and the like. On the other hand, sources of development failure define late comers to the game, believers in the superiority of government or collective ownership of the means of production, advocates of state planning and intervention, proponents of grievances of international exploitation and inequality, and such.[44]

To be sure, there is one inescapable reality. The struggle to overcome incapacity hence underdevelopment persists. The transition to modernity is indeed a time-consuming process. The road to sustained capacity innovation as history attests is filled with mistakes and disappointments, starts and stops, progression and regression, and the like. Failures make the wait for success even more intolerable, even more outrageous. But having a clearer view of the complexities of capacity innovation and a definitive path to attainment may be vital to rendering the waiting more tolerable. So then, the postulate extended in the opening stanza deserves repeating: investment in capacity innovation enables a scale of economic activity that prompts progressive and continuous structural transformation. NEPAD's efforts in encouraging and supporting the activities outlined in the path to capacity innovation may well lead to the noted endgame. Clearly only time will tell. But the prescribed pathway is not just food for thought but grounds for action as well.

This standalone exploration of African capacity innovation serves as an opening inquiry into a holistic approach to sustained African economic transformation. The following discourse in Chapter 4 peruses the talents, resources and efforts requisite to mounting a comprehensive capacity innovation strategy.

---

[44]David S. Landes, "Richard T. Ely Lecture," American Economic Review, Papers and Proceedings 80, No. 2 (May 1990): 1–13, in Gerald M. Meier, *Leading Issues in Economic Development*, New York: Oxford University Press, 1995, pp. 74–79.

4
Top Down–Bottom Up Capacity Innovation

Conventional intelligence on capacity innovation in the economic development domain underscores two approaches: "top down" where innovation is impelled by government action and/or the pursuits of other powerful entities; and "bottom up" where innovation is prompted by the "grassroots" – startup entrepreneurs, community organizations and other local entities. Somerville suggests an alternative approach: "side in" where innovation is driven by a partnership of community entities and influential individuals and/or organizations.[1] How these approaches take form in the real world is given to varying interpretations and narratives, highlighted by 'support for' and 'opposition to' arguments. Before setting forth the approach particular to this volume, it would be enlightening to outline at least a modicum of the perspectives on top-down and bottom-up innovation in application or in print.

To begin, consider the question raised by Conover: "Which path to job creation and prosperity do we prefer: A continued emphasis on trusting government with top-down economics, or greater emphasis on trusting consumers and entrepreneurs with trial-and-error bottom-up economics?"[2] He asserts that "Real top-down economics is the political equivalent of "intelligent design": Trusting the superior abilities of government experts to ensure that the right things happen and the wrong things don't" and supports his assertion by arguing that top-down prosperity strategy "requires faith that a president-appointed car czar can produce better results for auto-company bankruptcies than can a century of bankruptcy case law evolved from the bottom up. It requires faith that today's government regulators can predict—accurately and without bias—which energy technologies and corporations deserve taxpayer subsidies and which competing technologies and corporations government

---

[1] Peter Somerville, *Understanding Community: Politics, Policy and Practice*, University of Bristol: The Policy Press, 2016.

[2] Steve Conover, 'Top-Down' vs. 'Bottom-Up,' *The American*, September 2, 2012, https://www.aei.org/publication/top-down-vs-bottom-up/.

should therefore discourage or demonize. It requires faith that macro decisions by a few thousand government appointees can allocate healthcare better than micro decisions by millions of sufficiently insured healthcare consumers."[3] He concludes, quite emphatically, that "Top-down economics is government-knows-best economics."[4] Nee, initially avowing top-down economics but later questioning its ability to induce capacity innovation, submits that under the notion that the government – as the arbiter of the "rules of the game," is the primary source of capacity innovation, state-centered theory emphasizes the role of political actors.[5] Nee asserts that "the idea that politicians play a key role is substantively undeniable and intuitively appealing. With its comparative advantage in violence [political and economic control], the state operates as a monopolist that enjoys substantial cost advantages in institutional change."[6] Nee further asserts that "by contrast, a free-rider dilemma constrains the ability of economic actors to assume the cost of collective action to establish and enforce the rules of the game."[7] Thus it follows that, as Nee intimates, capacity innovation will come from political actors – the state, rather than economic actors – the stakeholders, since the latter would always face the "free-rider" problem. The state will continue to innovate to adjust to changing relative prices since it has no "free-rider" problem.[8] But, and now the disclaimer, the problem with the top-down approach – as Nee argues, is that it cannot explain the self-reinforcing, endogenous rise of private enterprise economies such as China and Vietnam, for example, which is the very foundation of capitalist economic development and the private ownership of the means of production driving private wealth accumulation.[9] Nee's argument is particularly noteworthy in China's case since "during the first decade of reform, the central government explicitly outlawed private enterprise as an ownership form in the transition economy. Reform leaders not only enforced rules against private enterprises, but predatory taxes and expropriation by local government of assets and wealth of peasant entrepreneurs highlighted the problem of insecure property rights for privately owned assets and wealth.

---

[3] Ibid.
[4] Ibid.
[5] Victor Nee, "Bottom-up Economic Development and the Role of the State" *CSES Working Paper Series*, Paper #49, Cornell University: Center for the Study of Economy & Society, January, 2010, p. 4, http://people.soc.cornell.edu/nee/pubs/bottomupecondevandstate.pdf.
[6] Ibid., pp. 4–5.
[7] Ibid., p. 5.
[8] Ibid.
[9] Ibid.

It was not until a decade after the start of economic reform, when the private enterprise was already growing rapidly, that the first constitutional amendment in 1988 eventually conferred legal status to private firms."[10] As for Vietnam, "in many ways the enormous economic and developmental gains . . . over the past quarter century owe more to "bottom-up" pressures for improved livelihoods and regional integration than to the "top-down" model of state capitalism . . ."[11]

On bottom-up capacity innovation Conover is decidedly empathetic, maintaining that "Real bottom-up economics is a system that emphasizes trust in the private sector to evolve organically, independently, and in desirable directions, within a tested and evolving legal framework. Bottom-up economics—a.k.a. "emergence" or "complexity economics"—cannot and does not dictate which technologies and firms will (or should) be the winners and losers; instead, it places heavy emphasis on trusting consumers and rule of law to sort them out—in the auto industry, for example. It trusts the private sector to evolve in the favorable direction of a higher aggregate standard of living, to allocate capital from those who have it to those who need it, to add new jobs that require new, higher-level skills, and to jettison or outsource obsolete jobs that require only yesterday's lower-level skills. It trusts adequately funded entrepreneurs to continue surprising the world with innovations rivaling those of the past and present—such as the Internet search algorithms of Google, the horizontal drilling technologies of Big Oil, the instant-communication platform of Twitter, the slicker-than-cash payment system of Square, and the low-cost mega supply chain of Wal-Mart. It trusts consumers to sort out winners from losers in a trial-and-error process."[12] Summing up, Conover proclaims that "Bottom-up economics is consumer-knows-best economics."[13] This narrative places Conover squarely on the side of bottom-up capacity innovation.

Ascani, Crescenzi and Iammarino offer further commentary on capacity innovation, approaching the matter from a regional development perspective. In so doing, they emphasize "the importance of a bottom-up approach to economic development given," as they subscribe, "the localized nature of this process and the frequent ineffectiveness of top-down policies" arguing

---

[10] Ibid.

[11] Yan Flint, "Capitalism Vietnamese-style: Combining Top-down with Bottom-up," *Rethinking Development in an Age of Scarcity and Uncertainty*, EADI/DSA, University of York, September 19–22, 2011, p. 4, http://eadi.org/gc2011/flint-625.pdf.

[12] Ibid.

[13] Ibid.

"that the increasing demand for decentralization of powers and resources from central governments to regional and local administrations in most parts of the world in the last decades can be interpreted as the acknowledgement that regional forces and characteristics are strongly relevant in shaping local development trajectories in a context of increasing globalization. In this framework, therefore, decentralization represents the capacity of heterogeneous regions and territories to tailor specific development strategies in order to address their particular needs and influence their own destinies. Hence, the relevance of the processes of decentralization is also connected to the shift from traditional top-down development strategies to bottom-up approaches by means of the increasing level of decision-making power that decentralization attaches to local authorities and institutions. In other words, decentralized governments are most likely to adopt regional development strategies where the evaluation of territorial strengths and weaknesses as well as the inclusiveness of local agents are at the core of policies."[14] In their view, "this makes innovation and development no longer a linear but a multidimensional process by affecting local relations, rules, absorptive capacity and the capability to re-use knowledge. Globalization sharpens the localized nature of innovation and development rather than alleviating it, since successful regions become able to exploit external knowledge as well as to serve international markets. Acknowledging that development is a localized process dependent on spatially-bounded elements as well as past trajectories provides an explanation for inequalities between regions within countries."[15] They contend "that the pattern of regional disparities is more evident in developing countries due to the scarcity of locations that are able to absorb external knowledge in these areas. Bottom-up policies are precisely designed to take into consideration forces that influence innovation and development in specific locations. Such policies are in contrast with traditional top-down strategies that basically offer the same general measures of economic policy regardless of local conditions and characteristics."[16]

Ascani et al., further assert that "the growing awareness of the importance of local features in shaping development trajectories crucially undermines one of the main characteristics of top-down policies: the transferability of 'universal' strategies to every region, regardless of local weaknesses and strengths.

---

[14]Andrea Ascani, Riccardo Crescenzi and Simona Iammarino, "Regional Economic Development: A Review," *Search Working Paper*, WP1/03, European Commission, January 2012, p. 3.
[15]Ascani, et al., ibid., p. 19.
[16]Ibid.

Such a 'one-size-fits-all' character of top-down policies has frequently led to the promotion of infrastructural investment as a way to improve market access of remote regions or endow poor areas with physical capital in the belief that returns from such a kind of investment are exceptionally high" and that "the attraction of large firms to locations with weak levels of industrialization has . . . led to the localization of firms within contexts which could not support the presence of industry mainly due to the lack of adequate skills and capabilities as well as institutional weaknesses."[17]

Extolling the efficiency of decentralized power and localized development initiatives, Ascani et al. assert that "decentralized administrations are empowered with the capacity to design and implement strategies that recognize the local cultural and socio-institutional underpinnings of regional economic interactions and behavior . . . which crucially differs from traditional top-down development strategies managed at the central level."[18] However, they affirm that decentralization of power may also cause inefficiency noting that "a mismatch between resources and authorities combined with a culture of soft budget constraint may be extremely inefficient. That is, in countries where devolved powers are larger than devolved resources, regions are increasingly dependent on external financing such as national transfers or bank loans. Moreover, as a result of soft budget constraint, information asymmetry in the form of moral-hazard may emerge. Consequently, regions may overspend and accumulate huge debts, being certain that the national government will bail them out. Hence, regional overspending may negatively affect the macroeconomic stability of the whole country potentially undermining macroeconomic stability."[19] Further, "central governments may be more efficient in the provision of goods and services because of economies of scale and scope. Indeed, the regional scale may be too small to be efficient and save costs."[20] It should be recognized that "a minimum critical mass is required to use powers effectively."[21] And, "territorial competition may reduce efficiency because it leads regions to incur costs related to making a location more attractive for external resources."[22] Moreover, decentralization, "may affect efficiency because of the possibility of overlapping competencies between

---

[17] Ibid., p. 11.
[18] Ibid., p. 13.
[19] Ibid., p. 15.
[20] Ibid.
[21] Ibid.
[22] Ibid.

different tiers of government that tend to replicate the same services."[23] It should be recognized as well that "efficiency is also reduced by the risk of corruption that is greater at local level where the interaction between economic and political agents is more frequent."[24]

Banks offers interesting ingress to the topic in sharing an approach to bottom-up capacity innovation in which the Nigerian government granted $60 million to 1,200 of its citizens for business startup and development. Over the course of three years, the initiative spawned hundreds more profitably operating new businesses, employing about 7,000 new workers. In light of this outcome, Banks poses the question: what if the donor community directed more funds to local entities – the grassroots and NGOs, and left the rest to entrepreneurial spirit and market forces? Further addressing the topic, several arguments are presented in support of expanded donor funding to local entities: locals better understand the needs of the local community and have a vested interest in developing meaningful solutions to local problems; larger (outside) organizations divert funds to operational budgets to sustain themselves; local organizations tend to be free of reporting requirements and therefore can be more focused on solutions; and grassroots grantees are more successful. Interestingly arguments are made, from the donors' perspective, for rejecting direct local funding: local NGOs don't have the necessary skills or capacity to meet proposal or reporting requirements; administering smaller grants suitable for smaller organizations is less cost-effective; mainstream grantees – such as larger NGOs, allow more efficient risk management; and donors are pressured to fund home country NGOs who then deliver funds and resources to host country locals.[25]

What is quite poignant here is that Banks' comments raise important questions about the efficacy of capacity innovation from the bottom: Is the objective to do "good" or to do "well?" Doing good is more of a moral issue couched in corporate social responsibility and socially responsible investing dogma whereas doing well is an economic issue couched in free enterprise doctrine and return on investment. Put another way: Doing good refers to equity and social contributions or returns to society. Doing well refers to efficiency and economic returns of the business (corporation). In other words, doing good means helping as many people as possible with available resources or doing what's beneficial for the most people within resource

---

[23]Ibid.
[24]Ibid.
[25]Ken Banks, "The Top-Down, Bottom-Up Development Challenge," *Stanford Social Innovation Review*, December 11, 2015.

constraints while doing well means generating profit or positive returns on investments or putting resources to their highest and best use.[26] Commenting on the notion of doing good and the social responsibility of the corporation, Friedman asserts that corporate officials serve no purpose beyond the interest of their stockholders. The view that they have a social responsibility "shows a fundamental misconception of the character and nature of a free economy. In such an economy, there is one and only one social responsibility of business – to use its resources and engage in activities designed to increase its profits so long as it stays within the rules of the game, which is to say, engages in open and free competition, without deception or fraud."[27] On the matter of doing well, Friedman's assertion on the responsibility of business when seen from a different perspective is instructive: when business succeeds in generating profit, it is doing well. Such is the case since increased profit reflects positive returns on investment. So the lesson here regarding the direction of donor funding is perhaps twofold. If donors value equity more highly, target the locals directly. On the other hand, if efficiency (cost-effectiveness) is of greater concern, mainstream grantees such as larger NGOs based in the home country should be used to deliver resources to locals. But decision making on this matter may be informed by Panda's assertion that NGOs employ bottom up and top down approaches to achieve their objectives.[28] Panda envisages bottom up as emphasizing "local decision making, community participation and grass-roots mobilization," and top down focusing on "lobbying and bargaining with the decision-making authorities such as government agencies, building up of pressures through various campaign mechanisms and advocacy activities."[29] Using an array of indicators, the bottom-up approach is functionalized through "awareness building efforts of NGOs, people's participation in different phases of projects, and people's involvement in creating people's institutions" and top down captures "NGOs' participation in advocacy activity, obtaining support

---

[26] Alejandrino J. Ferreria, "Do Good While Doing Well," http://www.philstar.com/business-usual/137504/do-good-while-doing-well, October 22, 2001. Alejandrino Ferreria is the associate dean of the Asian Center for Entrepreneurship (ACE) of the Asian Institute of Management.

[27] Friedman, op. cit., p. 112. See also Maria O'Brien Hylton, "Socially Responsible" Investing: Doing Good versus Doing Well in an Inefficient Market, *The American University Law Review*, Vol. 42, Issue 1, 1992, pp. 4–6. Biswambhar Panda, "Top Down or Bottom Up? A Study of Grassroots NGOs' Approach," *Journal of Health Management*, Vol. 9, Issue 2, May 2007, pp. 257–273.

[28] Biswambhar Panda, "Top Down or Bottom Up? A Study of Grassroots NGOs' Approach," *Journal of Health Management*, Vol. 9, Issue 2, May 2007, pp. 257–273.

[29] Ibid., p. 257.

from government authority and obtaining favorable court verdicts."[30] Panda finds that "despite the rhetoric . . . no grassroots NGO practices either a bottom-up or top-down approach exclusively."[31]

Further, in light of the issues raised in Chapter 3 concerning core challenges, Banks' question regarding local funding and market forces begs a further question: does bottom-up capacity innovation create market forces in such magnitude as to overcome the core challenges? Sanyal offers interesting perspective in *The Myth of Development from Below* that raises doubt in this regard. Focus on bottom-up development, Sanyal confers, stemmed from critics' claims that the top-down model had failed; their reasoning being that "the institutions created to foster development from the top had themselves become the greatest hindrance to development."[32] The brunt of the criticism was heaped upon the state but others were rebuked as well: "market institutions, such as large private firms, were criticized for taking advantage of various types of state protection which made them capital intensive and inefficient. Established political parties were criticized for seeking power by manipulation of the poor and collusion with the army and the elite. Trade unions were chastised for protecting the interests of only the "labor aristocracy," and for being incorporated by the state into "the system."[33] As replacement for the top-down paradigm's central objective of industrialization and economic growth, Sanyal instructs, was the bottom-up model with a different cast of actors, issues, values and modes of action, all of which were to collectively empower the poor.[34] "Instead of the state-administered, large-scale infrastructure projects that were central to the employment-generation strategy of the top-down model, it advocated small-scale, bottom-up projects that directly involved the urban and rural poor in income-generating schemes" which "were expected to generate profit, savings, and investment at the bottom, thereby eliminating the need for income to trickle down through the market hierarchy."[35] Participants received subsidized credit for initiating small enterprises which ranged from basket making and other low-skilled

---

[30] Ibid.
[31] Ibid.
[32] Bishwapriya Sanyal, *The Myth of Development from Below*, Department of Urban Studies and Planning, Massachusetts institute of Technology, p. 1, http://web.mit.edu/sanyal/www/articles/Myth%20of%20Dev.pdf.
[33] Ibid.
[34] Ibid.
[35] Ibid.

activities in rural areas to production of low-cost wage goods and provision of low-skilled services in urban areas.³⁶

Emergence of bottom-up capacity innovation, Sanyal informs, was to be assisted by the guiding hands of non-governmental organizations (NGOs) which are thought to be particularly appropriate for such an initiative as their modus operandi is community development and grassroots bonding. They are not profit-seeking and "their organizational priorities and procedures are diametrically opposite those of the institutions at the top."³⁷ Bottom-up projects have been implemented in Africa, Asia and Latin America. Sanyal personally conducted field work in Bangladesh and India where he investigated the workings of NGO-led bottom-up capacity innovation efforts. Sanyal's findings as well as those of other investigators are instructive. After several decades and thousands of bottom-up projects, Sanyal reports, "the economic impact of bottom-up efforts seems rather small and insignificant. These projects did not create new employment and income-earning opportunities for a large number of people; nor did they significantly increase the income of those fortunate few project beneficiaries. The barriers to success have been many. The main barrier, however, has been the lack of demand for the goods and services produced by the small businesses run by the poor . . . the bottom-up approach was deliberately designed to discourage economic linkages between the top and the bottom, because such linkages were seen as conduits for exploitative surplus extraction from the bottom to the top. . . . Contrary to the logic of development from below, economic stagnation at the top cannot be overcome by fostering small enterprises at the bottom; because the top and the bottom are linked . . ."³⁸ Sanyal further informs that the effectiveness of NGOs was thwarted by "their inability to cooperate with each other. This is particularly surprising since NGOs are thought to represent models of cooperation. In reality, though, the NGOs were extremely competitive and rarely formed institutional linkages among themselves. This was primarily due to the dependence on grants and donations which made every NGO claim that its particular organization was most effective in helping the poor. To support such claims, each NGO tried to demonstrate to the donor community how it alone had successfully designed and implemented innovative projects. . . . At best, their efforts created small, isolated projects that lacked the institutional support necessary for large-scale replication."³⁹

---

³⁶Ibid.
³⁷Ibid., p. 2.
³⁸Ibid., p. 8.
³⁹Ibid., p. 9.

Turning to the question asked earlier concerning whether bottom-up capacity innovation creates market forces in such magnitude as to overcome the core challenges, clearly the sheer magnitude of the core challenges requires the emergence of large-scale markets with the capacity to overcome problems associated with especially critical mass, comparative advantage, economic complementarity and economic integration. The bottom-up approach, certainly from Sanyal's perspective, appears to fall far short of redressing demand and supply constraints at the grassroots level to say nothing of its ability to prompt emergence of large-scale markets that are the focus of top-down capacity innovation initiatives. To this point, the Nigerian government's grant initiative presented by Banks raises similar concern. Although Nigeria with its rather sizeable population does not face the critical mass issue to the extent faced by most other African countries, what is the likelihood of replicating on a larger scale the small enterprises spawned by government grants? Importantly, can the Nigerian bottom-up model create comparative advantage sufficient to propel the country onto the world stage? Ultimately, this is the true test of a capacity innovation model.

By contrast, "China's post-Mao transition to a market economy," for example, offers important lessons in the context of bottom-up economics.[40] Coming on the thrust of bottom-up capacity innovation as extolled in *How China Became Capitalist*, China appears to have managed well the bottom-up approach to economic makeover. The authors, Coase and Wang, provide skillful in-depth narrative on China's economic transformation "from a deeply impoverished socialist country to the world's second-largest economy."[41] It is not all together clear, however, whether critical mass is the core reason for China's growth energized by attainment of comparative advantage in important areas of the economy, or whether emergent comparative advantage perhaps resulting from capital flows from the West bolstered by advantages of critical mass explains China's dramatic rise, or both. This is exceedingly important as a reversal of capital flows from the West – as signaled by, for example, the United States' "made-in-America" sentiment may upset the balance in the Chinese economy, thus bringing into question its bottom-up approach. This notwithstanding, Wolf's review of Coase and Wang's work grasps the essentials quite well and offers concise commentary on

---

[40] See Ronald Coase and Ning Wang, *How China Became Capitalist*, New York, NY: Palgrave Macmillan, 2012, p. 2.

[41] See Charles Wolf's commentary "A Truly Great Leap Forward" in *The Rand Blog*, http://www.rand.org/blog/2013/05/a-truly-great-leap-forward.html, Santa Monica, Calif.: The Rand Corporation, May 2, 2013. See also Coase and Wang, ibid.

their interpretation that "China's dramatic economic growth over the past three decades hasn't been a top-down process engineered by party leaders in Beijing. Instead, China's rise has been a bottom-up process driven by what the authors call the "four marginal revolutions."[42] Chronologically these entail the "household responsibility system" in agriculture that, the authors say, "emerged spontaneously in rural China" in the late 1970s and was implemented nationwide in 1982. This system allowed farmers to sell some of their output at free-market prices. The next "revolution" resulted from rural industry reform in the form of township and village enterprises, which impelled townships to behave like entrepreneurs in producing and marketing their products. These enterprises in turn led to the emergence of entrepreneurs and the "individual economy"—"a euphemism for private economy," the authors explain, that was intended to disguise the underlying capitalist reality of this third revolution. The fourth revolution was the establishment of the "special economic zones" in Shenzhen and several other towns in Guangdong and Fujian provinces, as well as in Shanghai and other coastal cities."[43]

The veracity of the "four marginal revolutions" is perhaps yet to fully unfold. The strength of China's bottom-up process very likely hinges on how well the grassroots are infused in the transformation, how extensively comparative advantage envelopes the whole of the economy, the independence and vibrancy of the non-state sector and its continued expansion, and ultimately whether the transformation can be sustained. The authors "insist that the future dynamism of China's economy depends on the continued growth of its private sector. But they are much less clear on the future of China's huge state-owned enterprises and of China's partial efforts to privatize them."[44] Nonetheless, the Chinese bottom-up capacity innovation process offers important insight on the various ways the process may take place.

An alternative view on capacity innovation may be gleaned from the perspective of "institutions" particularly in light of Easterly's question: "Is the bottom up view of institutions hopelessly pessimistic?"[45] He reckons that "It certainly undermines the naïve optimism implied by aid agency recommendations that institutions can be rapidly changed from the top

---

[42] Wolf, ibid.
[43] Wolf, ibid.
[44] Wolf, ibid.
[45] William Easterly, "Institutions: Top Down or Bottom Up?" *American Economic Review*: Papers & Proceedings 2008, 98:2, 95–99, http://www.aeaweb.org/articles.php?doi= 10.1257/aer.98.2.95

by political leaders. Even without a comprehensive theory of institutions, historical evidence, contemporary research, and common sense suggest that institutional change is gradual in the large majority of cases. Attempts at rapid, top down change can even have negative consequences. If that is reality, then an agenda of gradual reform that recognizes the constraints of bottom up evolution will lead to more hopeful results than a delusory top down attempt to leap to institutional perfection."[46] And he reasons that "The top down view also tends to go together with the view that there is one globally unique best set of institutions, toward which all societies are hopefully thought to be "developing." The development economist acts as a cross-country communicator of the institutions of the "advanced" society to the less informed in the "backward" society. The bottom up view of institutions is more open to the possibility that societies evolve different institutions even in the long run."[47] With a notion of resolving the rather dichotomous nature of top-down/bottom-up views, Easterly asserts that holding these two worldviews as opposing extremes is a caricature – "most views lie somewhere in between. The top down view is seldom advocated explicitly.... Nor is the most extreme bottom up view tenable, or we would not need formal states and laws at all, whereas in fact they are ubiquitous."[48] "The apparent effectiveness of top down formal institutions in rich societies may still depend on these institutions having evolved from the bottom up. If so, then attempting to introduce formal institutions into poor societies where bottom up factors are lacking will not replicate the institutional successes of rich countries."[49] Klien offers what is perhaps a capstoning critique, proposing that "instead of searching for the big idea, thinking that once we identify the correct one, we can simply unfurl it on the entire developing world like a picnic blanket, we should support local, incremental, experimental, attempts to improve social and economic wellbeing..." and he stresses that "international development is just such an invasive species. Why Dertu doesn't have a vaccination clinic, why Kenyan schoolkids can't read, it's a combination of culture, politics, history, laws, infrastructure, individuals—all of a society's component parts, their harmony and their discord, working as one organism. Introducing something foreign into that system—millions in donor cash, dozens of trained personnel

---

[46] Ibid.
[47] Ibid., p. 96.
[48] Ibid.
[49] Ibid.

and equipment, U.N. Land Rovers—causes it to adapt in ways you can't predict."[50]

4.1 Integrative Approach: Top Down–Bottom Up Convergence

Looking back, top down capacity innovation, though couched in deductive reasoning and macro-economic theory, is criticized because of government's power to impose development schemes, e.g., structural adjustment programs and arguably because of "the disappointing results of traditional top-down, supply-side sectoral development strategies in combating ... regional inequality...." particularly in the African context.[51] There appears to be empathy for bottom-up capacity innovation because it, as argued, represents the grassroots struggle against top down commands and allows grassroots input to local development schemes.[52] Bottom up capacity innovation – more advanced through inductive reasoning, takes place at the lower end of the development spectrum, particularly connected with grassroots populations. There is much to be said for employing the full range of talents and interests of the grassroots. Because they tend to be marginalized in the development process, their potential contribution to capacity innovation tends to be overlooked. But by being unable to develop their maximum productive potential, marginalization prevents their creation of much greater economic wealth than otherwise would be the case. This makes the entire capacity innovation process less efficient and the whole of the development process worse off than it would be without marginalization by imposing immense economic costs on nation building. The costs include the loss of national output as a consequence of holding grassroots populations below their maximum productive potential. They also include the loss of markets because the incomes of grassroots populations are kept low by economic exclusion, and the heavy social costs associated with

---

[50]Peter Klein, "Bottom-up Approaches to Economic Development," *Organizations and Markets*, November 19, 2014, https://organizationsandmarkets.com/2014/11/19/bottom-up-approaches-to-economic-development/.

[51]Andrés Rodríguez-Pose and Sylvia Tijmstra, *Local Economic Development as an Alternative Approach to Economic Development in Sub-Saharan Africa*, (A report for the World Bank), http://siteresources.worldbank.org/INTLED/Resources/339650-1440997189l4/AltOverview.pdf, p. 3.

[52]Ibid., p. 4.

efforts to remedy conditions connected with exclusion, such as substandard living conditions, crime, health challenges, and a host of other ills.[53]

Prompting capacity innovation particular to this volume requires a comprehensive engagement and realization of a range of economic activities approximating in some fashion those spelled out in the stages of economic growth, ladder of comparative advantage, core challenges and endgame philosophy. Such accomplishments are informed by the preceding narrative on top down – bottom up capacity innovation and involve an assault on economic marginalization and underdevelopment through marshalling and operationalizing efforts and talents of government, highly advanced organizations and grassroots as well. These call for structural development and skillful resource allocation from the top of the development spectrum forging a path down to the grassroots and from the bottom of the spectrum pushing upward to interfacing with top-down activity. This top down–bottom up convergence essentially identifies with Crescenci and Rodríguez-Pose's notion of a 'meso-level' scheme whereby top down capacity innovation links up with bottom up capacity innovation in a chain of economic activity necessary to induce a more efficient holistic development process.[54] The 'meso-level' orientation stems from Crescenci and Rodríguez-Pose's concern that top down and bottom up capacity innovation policies have been generally promoted as two irreconcilable ends of the development spectrum: "Top-down policies, solidly based in micro- and macroeconomic theories, but lacking the adequate flexibility and 'place-awareness' to respond to local complexity; bottom-up approaches much more responsive to diverse territorial needs,"[55] but couched in inductive reasoning.[56] This division, they assert, need not remain static and "the foundations of top-down and bottom-up development policies

---

[53]Ibid. See Benjamin F. Bobo, *Locked In and Locked Out: The Impact of Urban Land Use Policy and Market Forces on African Americans*, Westport, Connecticut: Praeger Publishers, 2001, p. 90; and U.S. Commission on Civil Rights, "White Gains and Social Costs from Subordination of Blacks," in Harold G. Vatter and Thomas Palm, eds., *The Economics of Black America*, New York: Harcourt, Brace, Jovanovich, Inc., 1972, pp. 144–148.

[54]Riccardo Crescenzi and Andrés Rodríguez-Pose, "Reconciling Top-Down and Bottom-Up Development Policies," *Working Paper Series in Economic and Social Sciences 2011/03*, IMDEA Social Sciences Institute, January 2011 (The comments made herein reference this version of the paper.) See also Riccardo Crescenzi and Andrés Rodríguez-Pose, "Reconciling Top-Down and Bottom-Up Development Policies," *Environment and Planning A*, 43(4), 2011, pp. 773–780. ISSN 0308-518X.

[55]Ibid., p. 1.

[56]Ibid., p. 2.

can be reconciled in a joint 'meso-level' conceptual framework which can serve simultaneously as a deductive justification for bottom-up local and regional development policies and as a coordination device between different policies."[57] To this affirmation, Crescenci and Rodríguez-Pose submit that "On the one hand, the literature on local and regional development has developed sound 'meso-level' analytical tools which combine inductive and deductive perspectives on local and regional development dynamics. On the other, the macro-economic approach to development has made significant steps towards becoming more open to inductive reasoning and, hence, to the consideration of local specificities."[58] An integrative approach, as this commentary avows, provides a common conceptual underpinning for both top-down and bottom-up policies.[59] Further, regional development literature holds a number of works that explicitly provide analytical support (both qualitative and quantitative) for local development policies by comparing their strategies, performance and processes in a systematic fashion. These analyses give clear credence to an integrative structure as rationalizing top down-bottom up convergence.[60]

4.2 Capacity Innovation for Prosperity: Top Down–Bottom Up with a Twist

In light of the integrative-convergence notion, this volume proposes a top down-bottom up integrative framework – but with a twist, as demonstrated in Exhibit 4.1, involving the government (NEPAD) linking up with the private sector (MNC) – the twist – in a cross-fertilizing fashion to, in effect, stimulate market forces from the top – essentially mimicking top-bottom convergence, and community input (grassroots) partnering with professional talent (NGOs) in a cross-fertilizing fashion to prompt market forces from the bottom, with both processes creating an operating framework that allows the entities to function in a holistically cross-fertilizing manner – as the forces converge, and to accrue synergies emanating therefrom. As the dominant entity, the NEPAD/MNC arrangement encourages and supports the grassroots/NGO

---

[57] Ibid., p. 1.
[58] Ibid., p. 3.
[59] Ibid., p. 5.
[60] Ibid., p. 7. The capability of these analyses to support policy making has limitations related to discrete underlying 'benchmarking.'

72 Top Down–Bottom Up Capacity Innovation

Exhibit 4.1 Top Down-Bottom Up Cross Fertilization (Framework of Coordination and Convergence).

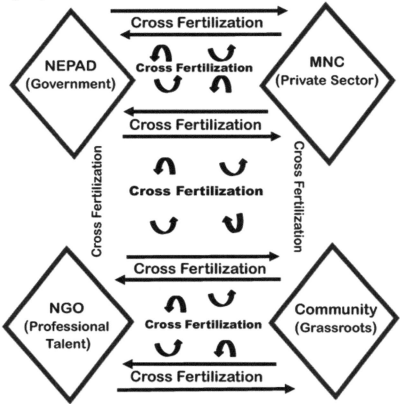

arrangement to promote more efficient comprehensive cross-fertilization. (This framework is presented in Chapter 7.) With regards to Nee's concern that the top-down approach cannot explain the self-reinforcing, endogenous rise of private enterprise economies, this is less of an issue in the proposed integrative framework as the top-down actor (NEPAD) is joined with the private enterprise actor (MNC) to induce the major influence on capacity innovation with the NGO and grassroots to influence economic transformation from the bottom. The integrative, convergence and cross-fertilization process ultimately promotes "take-off" as enunciated in Chapter 3.

The cross-fertilization scheme reckons with Crescenci and Rodríguez-Pose's view that "By cross-fertilizing theories rather than simply linking-up indicators ... clear, measurable, theoretically-grounded targets for local and regional policies can be identified."[61] It aligns with their joint 'meso-level' conceptual framework in incorporating coordination and convergence features in the top down-bottom up operating framework between NEPAD/MNC and Grassroots/NGO. [62] Further, it recognizes the benefits that may accrue from a combination of top-down and bottom-up capacity innovation such as that pursued by Vietnam in the "doi moi" (renovation) economic renewal program that focused on foreign direct investment and market pricing. Since the end of the Vietnam conflict, the country has undergone economic transformation from famine and deprivation to the second largest supplier of clothes to the United States, the world's second-largest rice and coffee exporter and to a Human Development Index (HDI) ranking of .590 in 1975 to .666 in 2014.[63] The proposed framework is also in concert with Hosman and Fife's notion that understanding the nexus of where bottom-up needs and innovation meet top-down capabilities and funding can address sustainability issues, and create a feedback loop between needs identification and technological expertise, and further that, MNCs interested in engaging the grassroots may be wise to consider projects that address societal as well as economic concerns; simply treating the grassroots as consumers ignores the human development paradigm's notion that people have a fundamental right to a range of life choices that lead to enhanced capability and empowerment, to increased incomes and purchasing power, and furthermore, to redress of the needs of the "missing middle" – the underdeveloped small and medium enterprise sector across the African continent.[64]

Particular attention is accorded apprehensions about the attraction of large firms to locations with weak levels of industrialization or which could not support the presence of industry owing to institutional weaknesses, the lack of adequate skills and competences or the inability to absorb external

---

[61] Ibid. p. 8.

[62] Ibid. p. 8.

[63] Flint, ibid., p. 18 and, Human Development Reports, United Nations Development Program, 2015, http://hdr.undp.org/en/composite/HDI, and Human Development Index 1975–2005 – Country Rankings, 2008.

[64] Laura Hosman and Elizabeth Fife, The Use of Mobile Phones for Development in Africa: Top-Down-Meets-Bottom-Up Partnering, *The Journal of Community Informatics*, Vol. 8, No. 3, 2012, pp. 1–5, http://ci-journal.net/index.php/ciej/article/view/560/938. For a discussion of the Human Development Paradigm, see Benjamin F. Bobo, *Rich Country, Poor Country: The Multinational as Change Agent*, Westport, CT: Praeger, 2005, pp. 200–201. For original discourse on the human development paradigm, see Mahbub ul Haq, *Reflections on Human Development*, Oxford: Oxford University Press, 1995.

knowledge. Not overlooked here is the importance of local features in shaping development trajectories and that capacity innovation frameworks may not be transferable from region to region regardless of local weaknesses and strengths.[65] The framework recognizes as well the South African experience with local economic development projects – largely community development oriented; many projects failed to live up to expectations. Intervening issues related to internal constraints, absence of local resources and skills, and inability to attract significant FDI by the country as a whole are among the primary causes of failure. In light of South Africa's more advanced position relative to most African countries particularly in terms of GDP per capita and transport infrastructure, their unfavorable experience with local development initiatives should be seriously heeded in designing future capacity innovation frameworks.[66] And, the NEPAD/MNC-Grassroots/NGO operating framework incorporates the local economic development concept as "essentially a process in which local governments and/or community based groups manage their existing resources and enter into partnership arrangements with the private sector, or with each other, to create new jobs and stimulate economic activity in an economic area." [67]

In the proposed framework, there are essentially four actors: New Partnership for Africa's Development (NEPAD), multinational corporation (MNC), grassroots (community) and non-governmental organization (NGO). It constitutes a well-rounded structure predicated upon inclusiveness: all actors are essential entities and have important roles in occasioning the cross-fertilization and capacity innovation process. Synergistically, the multinational corporation is the key actor in the framework. Regarded by Bobo as the "change agent" in the pursuit of sustained development transformation by underdeveloped countries, the multinational corporation has the power and knowhow to induce the scale of capacity innovation necessary to achieve such objective. Reflecting on the development process in his book, *Rich Country, Poor Country: The Multinational as Change Agent*, Bobo asserts that "the inability to induce sustainable change may be due more to application of an ineffective

---

[65] Ascani, et al., op. cit., pp. 11 and 19.
[66] Rodríguez-Pose and Tijmstra, op. cit., p. 6.
[67] Andrés Rodríguez-Pose and Sylvia Tijmstra, *Local Economic Development as an Alternative Approach to Economic Development in Sub-Saharan Africa*, (A report for the World Bank), 2005, p. 3, http://siteresources.worldbank.org/INTLED/Resources/339650-1144099718914/AltOverview.pdf. See also M. Zaaier, M. and L. M. Sara (1993). "Local Economic Development as an Instrument for Urban Poverty Alleviation: A Case from Lima, Peru." *Third World Planning Review* 15, 1993, p. 129.

change model than lack of effort ... the multinational is viewed as a change agent with the expertise and, given the appropriate operating framework, the capacity to ... level the playing field between rich and poor countries. In the process, poor countries would develop some measure of comparative advantage. At that point ... these countries would be in a position to further develop comparative advantage under their own powers and capacities."[68] Presumably, in the interest of sustaining accruing synergies, a continuous relationship with MNCs and NGOs would be encouraged. Sustained transformation relies on comparative advantage emanating, in the context of the proposed framework, from top-down capacity innovation through investment in larger-scale economic activity/development projects integrated with foreign direct investment and capital diffusion – top down–bottom up with a twist, and from bottom-up activity through investment in smaller-scale initiatives integrated with NGO input. In the context of foreign direct investment, the multinational corporation is the capital provider and responsible for all phases of FDI project initiation, implementation and technology transfer. Ultimately transmission, absorption and assimilation of technology promotes technology transfer to the host country; a process that obtains capacity innovation and highlights the change agent capability of the multinational corporation. The non-governmental organization plays an important role in technology transfer as well; providing technical assistance to grassroots, community organizations and entrepreneurs that enable these parties to more efficiently conduct local initiatives.

4.3 Land Tenure, Property Rights and Ownership

Discourse such as the foregoing would be incomplete without at least some narrative on land property rights. It may be affirmed that land is power; land ownership is economic power; land ownership by the grassroots is absolute economic power – or is it? This is important particularly in the African domain where customary or communal tenure arrangements exist and where land ownership has been much debated as having a key role in capacity innovation. Who owns what and what proof exists are of major concern. If there were definitive answers to these questions, would they matter? Would land titling/registration under the formal land tenure system – a top down approach to resolve uncertainty over property rights, have any effect on the incentive

---

[68] See Benjamin F. Bobo, *Rich Country, Poor Country: The Multinational as Change Agent*, Westport, CT: Praeger, 2005, pp. xx and 149.

to invest? After all, is it not the incentive to invest or to engage in productive entrepreneurial activity that drives optimal outcomes? Holding land as dead capital or employing it in less than its highest and best use would result in less than optimal outcomes. From the standpoint of economic analysis, land is a scarce factor of production and would command a high produce price, despite the system of land tenure.[69] "The most effective way to stimulate productive entrepreneurial activity is to diminish relative rewards to unproductive or destructive rent-seeking and increase payoffs to productive entrepreneurial activity."[70] When considering the effects of land tenure arrangements on economic development, the investment aspect must be weighed.[71]

Top down land reform has been justified based on assertion that confused land title, implied by the "bundle of rights theory" – the right to use, sell, lease or give away property, raises uncertainty about land ownership hence raises the specter of destructive rent-seeking. While this would appear to pejoratively impact incentive to invest, stern criticism of top-down behavior contends that: "The goal of these approaches is noble as scholars have argued for decades that uncertainty over land titles is a major obstacle to development. However, when it comes to implementation, the top down approaches these scholars and policy-makers have favored have failed to produce much."[72] Importantly, various empirical studies report a modest effect of land titles on the incentive to invest; such was the result of a study of Kenyan farmers, for example.[73] Furthermore, land titling may have expanded rather than reduced the opportunistic behavior that institutions are thought to restrain. In some instances sellers pledged land as collateral for loans but failed to report it to buyers. Some sellers engaged in the same land transactions with several buyers at once.[74] Other studies in Africa produced similar results: a study of land titles

---

[69] Gerald M. Meier, *Leading Issues in Economic Development*, Sixth Edition, New York: Oxford Ubiversity Press, 1995, pp. 419–421.

[70] Nee, op. cit., p. 8.

[71] Meier, op. cit., p. 421.

[72] See Kipaya Kapiga, "Bottom-up and Top-down Approaches to Development," *Global Social Entrepreneurship*, http://global_se.scotblogs.wooster.edu/2011/06/26/bottom-up-and-top-down-approaches-to-development/.

[73] See William Easterly, "Institutions: Top Down or Bottom Up?" Design and Reform of Institutions in LDCS and Transition Economies, *American Economic Review*: Papers & Proceedings 2008, 98:2, p. 97, http://www.aeaweb.org/articles.php?doi=10.1257/aer.98.2.95. Migot-Adholla, Shem E., and Frank Place. 1998. "The Economic Effects of Land Registration on Smallholder Farms in Kenya: Evidence from Nyeri and Kakamega District." *Land Economics,* 74(3): 360–73.

[74] Ibid., pp. 97–98.

4.3 Land Tenure, Property Rights and Ownership 77

in Burkina Faso found no effect of land titles on incentive to invest; a land titling program in Cameroon failed to consolidate individual property rights; land titling of plot-specific investment in rice fields in Madagascar showed no effect; and a summary of land titles in Africa showed essentially no effect of titles on investment or access to credit.[75] Further, study results evidence that indigenous property systems in Africa do not constrain investments in increased land productivity, and that there is little benefit of formal land rights compared to indigenous systems.[76]

In light of the range of study results, Easterly asserts that "After all this research and experience, the aid donors today remain stuck on some kind of idealized comprehensive (top down) government reform that would somehow make formal registration of land titles "optimal" and as evidence of such notion, Easterly notes that "The United Nations Millennium Project . . . said, for example: "The rule of law involves security in private property and tenure rights . . . upholding the rule of law requires institutions for government accountability . . . this requires a well functioning and adequately paid civil service and judiciary, proper information technology (for registration of property. . . .)"" and that "The World Bank . . . *Land Policies for Growth and Poverty Reduction* . . . concludes. . . . "The establishment of a land policy framework to guide the sequencing of specific interventions in the sector can have multiple benefits in generating consensus, helping to prioritize actions, and (by ensuring participation in the implementation and monitoring of these interventions) avoiding costly errors. Given the long-term nature of interventions in the area of land policy . . . integration into the broader development strategy is particularly relevant to provide a basis for relating land policy to other interventions. . . .;" to this Easterly laments "Here, the donors' answer is more computers to register formal land titles!""[77]

In support of its position on top-down land policy and government land registration, The World Bank sponsored a major policy development initiative that culminated in the book, *Securing Africa's Land for Shared Prosperity: A Program to Scale Up Reforms and Investments*, authored by Frank F. K. Byamugisha. The work exhibits a decidedly different perspective on top-down land titling policy in Africa than shown in the foregoing. The view taken is that "The region's poor record of development suggests that it has not leveraged its abundant agricultural land . . . to generate shared and

---

[75] Ibid., p. 98.
[76] Ibid.
[77] Ibid.

sustained growth or eradicate poverty. There is a general disconnect between abundant land and development: ... Africa has the most land available and suitable for agriculture, the highest productivity gap among regions ... and the highest poverty rate. . . . Urban slums have proliferated, undermining urban development and poverty reduction efforts."[78] Africa is home to nearly half of the world's usable uncultivated land, yet as Byamugisha asserts, "despite its abundant agricultural land . . . Africa is still mostly poor and has been unable to translate its recent robust growth into rapid poverty reduction. These examples suggest that poor land governance—the manner in which land rights are defined and administered—may be the root of the problem. . . . Despite the determination and effort of leaders to improve land governance, many challenges exist. The following are the most binding: *Land grabs*. Investors have claimed millions of hectares . . . and poor governance leads to violations of the principles for responsible agro-investment and dispossession of local communities. . . . *Land vulnerability*. Only about 10 percent of rural land in . . . Africa is registered; the rest is undocumented, informally administered, and thus vulnerable to land grabbing and expropriation without adequate compensation. . . . *Inefficient land administration*. It takes twice as long and costs twice as much to transfer land in . . . Africa as in Organization for Economic Cooperation and Development (OECD) countries. . . . *Corruption*. According to a study in 61 countries by the Food and Agriculture Organization (FAO) and Transparency International, weak governance increases the likelihood of corruption in land administration. . . . *Low capacity and demand for professionals*. Ghana, Kenya, and Uganda each have fewer than 10 professional land surveyors per 1 million population versus Malaysia (197) and Sri Lanka (150) . . ."[79]

There are clearly opposing views on the import of land titling and registration; the central issue being whether uncertain land title has a negative impact on incentive to invest. While evidence may support the notion that there is little benefit of formal land rights, hence land titling essentially does not impact incentive to invest, the pertinent question may be: who is doing the investing? Since land grabs by investors refers to major investors of the FDI kind, land titling may not necessarily impact incentive to invest, certainly not to any measurable extent and certainly not to major investors. The sheer

---

[78] Byamugisha, Frank F. K. 2013. *Securing Africa's Land for Shared Prosperity: A Program to Scale Up Reforms and Investments*. Africa Development Forum series. Washington, DC: World Bank. doi: 10.1596/978-0-8213-9810-4. License: Creative Commons Attribution CC BY 3.0, p. 1.

[79] Ibid., pp. 1–2.

4.3 Land Tenure, Property Rights and Ownership

power of major investors simply allows them to impose their will on the property system and thereby dictate outcomes; hence their incentive to invest is not compromised by the land system's structure. As the finding that formal land rights provide little benefit to investors does not set apart or distinguish between major investors and smaller indigenous land buyers, acquisition of large tracts of land by foreign investors overshadows the experience of indigenous investors; they may be indeed dissuaded by uncertain land title but the finding simply does not capture the effect because of the rather insignificant clout of smaller investors.

Investment in African land appears to be in the midst of a paradigm shift. In *Land Grabs in a Green African Economy: Implications for Trade, Investment and Development Policy*,[80] Odari informs that "...the emerging trend of increased foreign direct investment (FDI) in land in Africa, ...has been popularly christened as the new 'land grab' on the continent. This is a consequence of the acquisition of vast tracts of land by foreign investors, which has been generating considerable criticism. It has been argued that the phenomenon is simply an ill wind that bears no good for the continent. However, it also represents an opportunity to address perennial food shortages by putting under-utilized land to use through large-scale agricultural production, as well as injecting much-needed FDI into the continent."[81] Odari "recommends reforms to the legal, policy and regulatory frameworks, as well as the adoption of sustainable business models that can promote"[82] efficient and equitable investment in African land; the objective being to encourage development of a more disciplined market structure while at the same time recognizing the troubling experience of indigenous stakeholders. To be sure, the experience of indigenous stakeholders must be heeded lest 'an ill wind that bears no good for the continent' is to become a storm across Africa.

But perhaps this need not be the case as there are a number of instances in which countries have enhanced land governance: "in 1978, China dismantled collective farms and used long-term leases to confer land rights on households, which launched an era of prolonged agricultural growth that transformed rural China and led to the largest reduction in poverty

---

[80]Godwell Nhamo and Caiphas Chekwoti (eds), *Land Grabs in a Green African Economy: Implications for Trade, Investment and Development Policy*, Pretoria, South Africa: Africa Institute of South Africa, 2014.

[81]Edgar Jalang'o Odari, "Africa's 'Agrarian' Revolution: Land and Policy Prescriptions to Promote Impact Investment in Foreign Acquisitions" in Nhamo and Chekwoti, ibid., Chapter 2, p. 12.

[82]Ibid.

in history. In Argentina, Indonesia, and the Philippines, legal recognition of land rights for residents in slum areas have improved the quality of their housing and the value of their plots."[83] For the African continent, Byamugisha has proposed a multi-point program to address land titling and land administration issues including establishing boundaries and registering communal rights; initiating systematic titling; eliminating restrictions on land rental markets; upgrading and computerizing land information systems; observing the principle of eminent domain; linking land use planning to a national land policy; and developing a sound tax policy, and a computerized tax assessment and tax collection system.[84] He advises that: "unless communal and individual land rights are registered and land governance is improved, the recent surge in foreign direct investment in . . . Africa will not generate shared and sustained growth, as disruptions will likely arise from the dispossession of local communities, and investors' deals will face severe uncertainty or collapse."[85]

Clearly the interests of major investors as well as smaller investors are at the heart of African land tenure debate. Major investors – typically the source of FDI, are highly prized in capacity innovation; capital diffusion (investment) by these actors may command widespread land use for major business initiative. This activity may provide massive inducement to economic capacity driven by shareholder wealth maximizing behavior with potentially very significant benefit accruing to indigenous stakeholders. At the same time, indigenous stakeholders – particularly grassroots, engage in land use (investment) that may spawn economic activity as well. Expectations are that these actions take place efficiently – for purpose of optimal outcomes but equitably – for purpose of fairness. Concern arises in an efficiency-equity tradeoff; it should not be viewed by indigenous stakeholders as an interdiction of their economic power, that is, marginalization should not be the perceived outcome. Thus, the business model should intensely recognize the expectations of indigenous stakeholders. The experience of Madagascar should be a constant reminder of the risk of ignoring the concerns of indigenous stakeholders. In this instance, a military coup resulted from citizen unrest surrounding the large-scale selling of agricultural lands that reached a tipping point in 2008 fueled by a government land deal with South Korean multinational Daewoo

---

[83] Byamugisha, op. cit., p. xvi.
[84] Ibid., pp. 5–6.
[85] Ibid., p. 36.

Ltd.[86] Certainly this serves as backdrop to the Africa-MNC strategic alliance framework proposed in this volume and informs the discourse throughout that African stakeholder interest is at the heart of market impacts and institutional changes suggested by and endemic to the notions advanced herein.

---

[86]Benjamin D. Neimark, "The Land of Our Ancestors: Property Rights, Social Resistance, and Alternatives to Land Grabbing in Madagascar," LDPI Working Paper 26, The Land Deal Politics Initiative, www.iss.nl/ldpi landpolitics@gmail.com, March 2013, p. 1.

5

The Country Capacity ID

A country's traversal of the ladder of comparative, presented in Chapter 3, is impacted by its capacity ID. The country capacity ID, as Exhibit 5.1 conveys, is a means of identifying the range of internal resources available to countries from which a range of capacity innovation activity is spawned. A strong capacity ID derives from power-dominant attributes upscale characteristics: sizeable population as basis for and giving rise to critical mass; bountiful natural resources of commercial value from which to derive foreign exchange as well as support development activity; high literacy suggesting well educated masses and presupposing advanced skills and creative talent; high human development index score signaling enlarged human choices and richness of human lives; capital formation and infrastructure development and readiness suggesting developed markets, comparative advantage and global competitiveness; foundational institutions and citizen participation apparatus suggesting core institutional mapping and government structure essential to supporting and encouraging good governance and rule of law; foreign direct investment indicating external interest and confidence in local investment opportunities important to spawning a range of development activity; and military resources to provide for national security, support law enforcement and protect foreign direct investment.

On the other hand, a weak capacity ID suggests serious deficits in human, economic and political wherewithal: population too small to achieve critical mass thus unable to accommodate enterprise of any efficient scale; little or no natural resources of commercial value thereby limiting foreign exchange and development opportunity; low literacy and human development index scores suggesting poorly educated masses, retarded skills and creativity as well as constrained human choices and quality of life; weak capital formation and infrastructure development suggesting limited investment activity and economic diversification giving rise to fragile market structure and restricted trade opportunity; weak institutional mapping and citizen participation

Exhibit 5.1 Country Capacity ID.

Power-Dominant Attributes
Critical mass population
Natural resource abundance
Highly literate population
Strong capital formation
Developed infrastructure
Good institutional mapping
Observance of rule of law
Citizen participation apparatus
Strong market structure
Significant FDI
Strong military resources

Weak-Deficient Attributes
Mini-state population
Natural resource deficit
Highly illiterate population
Weak capital formation
Underdeveloped infrastructure
Poor institutional mapping
Abuse of rule of law
Limited citizen participation
Weak market structure
Insignificant FDI
Weak military resources

apparatus occasioning self-serving political will, flawed governance and abuse of rule of law; foreign direct investment of minor significance suggesting limited local investment opportunity or inability to attract external interest in local investment; and absence of military forces to provide for national defense, support local policing efforts and protect foreign direct investment.

All-in-all, capacity ID attributes are central to development as they interface and interact causally as basic forces in establishing a process of capacity innovation that promotes modernity and the progression of the socioeconomic system. In essence, as Gerald Meier observes, "To interpret development in terms of a process involving causal relationships should prove more meaningful than merely identifying development with a set of conditions or a catalog of characteristics."[1] Gunnar Myrdal asserts that development is the "upward movement of the entire social system" or attainment of "ideals

---

[1]Gerald M. Meier, *Leading Issues in Economic Development*, Sixth Edition, New York: Oxford University Press, 1995, p. 7.

of modernization."[2] Logically, progression of the socioeconomic system may be interpreted as successful capacity innovation.

Clearly a strong capacity ID enables the capacity innovation process, but a weak capacity ID effectively relegates a nation to a 'locked in and locked out' position in the context of local, regional and indeed global trade interaction and economic competitiveness. Serious deficits in core attributes, the life blood of the capacity ID, have gravely frustrated the efforts of Africans to improve internal capacity by constraining development options, thus impeding not only their ability to develop markets outside the continent but their ability to develop local markets as well. Bobo refers to this as the "locking effect" – whereby African nations are locked into poorly developed local markets by virtue of a weak capacity ID and locked out of global markets by virtue of the same condition.[3]

Every country has a capacity ID, expressed through a range of core attributes that form the foundation of economic strength. There are perhaps common core attributes among countries that make possible a comparative assessment of relative strengths and weaknesses and moreover a general assessment of a region's capacity for development. Clearly, there is some danger in suggesting common attributes of countries since their characteristics may be far ranging and quite diverse. Nevertheless, it is useful to attempt some notion of 'comparability' or 'generality' if only to suggest a simple expression of commonality, hence understanding of cause and effect. This is important when considering a country's ability to contribute to redressing particularly structural problems, both local and regional, associated with matters such as mini-states, critical mass, economic complementarity, comparative advantage and economic integration that are central to African capacity innovation concerns. Exhibit 5.2 identifies core attributes that are critically important to capacity innovation and provides a country-by-country comparative framework affording a detailed look at core capacity attributes across the entire spectrum of African countries. This framework offers insight into a country's, and in an aggregate context a region's, dominant or deficient core attributes, thus aids in judging whether African countries have strong or weak core capacity IDs, hence the African region in general. This outcome provides

---

[2]Gunnar Myrdal, Asian Drama, 1968, p. 1869, as quoted in Gerald M. Meier, ibid. See also Chapter 3 in this volume for modernization narrative.

[3]See Benjamin F. Bobo, *Locked In and Locked Out: The Impact of Urban Land Use Policy and Market Forces on African Americans*, Westport, Connecticut: Praeger Publishers, 2001, p. 6.

Exhibit 5.2 African Countries Core Capacity Attributes.

Country	Population (million)	Resources of Commercial Importance	Literacy Percentage (age 15 and over can read & write)	Human Development Index (HDI)	Capital Formation (investment) % Growth 2014	Infra-structure*	Institutions – Rule of Law	Citizen Participation	FDI 2014 ($ billion)	Military/[Military Spending Budget Per Capita – through 2015 ($US)]
Algeria	39.67	petroleum, natural gas, iron ore, phosphates, uranium, lead, zinc	69.9	0.736	6.4	ND/Poor	Parliament/ Constitution/ Branches of Government	Republic/ multi-party/ suffrage	1.505	Army, Navy, Air Force [266]
Angola	25.02	petroleum, diamonds, iron ore, phosphates, copper, feldspar, gold, bauxite, uranium	67.4	0.532	NA	ND/Poor	Parliament/ Constitution/ Branches of Government	Republic/ multi-party/ suffrage	1.922	Army, Navy, Air Force [166]
Benin	10.88	small offshore oil deposits, limestone, marble, timber	34.7	0.480	0	ND/Poor	Parliament/ Constitution/ Branches of Government	Republic/ multi-party/ suffrage	0.377	Army, Navy, Air Force [6]
Botswana	2.26	diamonds, copper, nickel, salt, soda ash, potash, coal, iron ore, silver	81.2	0.698	16.7	ND/Poor	Parliament/ Constitution/ Branches of Government	Republic/ multi-party/ suffrage	0.393	Army (with Air wing) [179]
Burkina Faso	18.11	manganese, limestone, marble; small deposits of gold, phosphates, pumice, salt	21.8	0.402	1.1	ND/Poor	Parliament/ Constitution/ Branches of Government	Republic/ multi-party/ suffrage	0.341	Army, Air Force, National Gendarmerie [6]

The Country Capacity ID 87

Country		Resources							Military
Burundi	11.18	nickel, uranium, rare earth oxides, peat, cobalt, copper, platinum, vanadium, arable land, hydropower, niobium, tantalum, gold, tin, tungsten, kaolin, limestone	59.3	0.400	10.5	ND/Poor	Parliament/Constitution/Branches of Government	Republic/multi-party/suffrage	N/A	Army (with air wing) [7]
Cameroon	23.35	petroleum, bauxite, iron ore, timber, hydropower	67.9	0.512	13.5	ND/Poor	Parliament/Constitution/Branches of Government	Republic/multi-party/suffrage	0.501	Army, Navy, Air Force [16]
Cape Verde	0.52	salt, basalt rock, limestone, kaolin, fish, clay, gypsum	76.6	0.646	NA	ND/Poor	Parliament/Constitution/Branches of Government	Republic/multi-party/suffrage	0.132	Army, Coast Guard [16]
Central African Republic	4.90	diamonds, uranium, timber, gold, oil, hydropower	51.0	0.350	18.7	ND/Poor	Parliament/Constitution/Branches of Government	Republic/multi-party/suffrage	0.003	Army, Air Force [4]
Chad	14.04	petroleum, uranium, natron, kaolin, fish (Lake Chad), gold, limestone, sand and gravel, salt	47.5	0.392	NA	ND/Poor	Parliament/Constitution/Branches of Government	Republic/multi-party/suffrage	0.760	Army, Air Force [9]
Comoros	0.79	NEGL	57.0	0.503	−8.1	ND/Poor	Parliament/Constitution/Branches of Government	Republic/multi-party/suffrage	0.014	Army [15]

(Continued)

88 The Country Capacity ID

Exhibit 5.2 Continued

Country	Population (million)	Resources of Commercial Importance	Literacy Percentage (age 15 and over can read & write)	Human Development Index (HDI)	Capital Formation (investment) % Growth 2014	Infra-structure*	Institutions – Rule of Law	Citizen Participation	FDI 2014 ($ billion)	Military/[Military Spending Budget Per Capita – through 2015 ($US)]
Congo, Democratic Republic	72.27	niobium, tantalum, petroleum, industrial and gem diamonds, gold, silver, zinc, manganese, tin, uranium, coal, hydropower, timber	66.0	0.433	9	ND/Poor	Parliament/ Constitution/ Branches of Government	Republic/ multi-party/ suffrage	(0.343)	Army, Navy, Air Force [2]
Congo, Republic of	4.62	petroleum, timber, potash, lead, zinc, uranium, copper, phosphates, gold, magnesium, natural gas, hydropower	84.0	0.591	22.1	ND/Poor	Parliament/ Constitution/ Branches of Government	Republic/ multi-party/ suffrage	5.505	Army, Navy, Air Force [30]
Cote d'Ivoire	22.70	petroleum, natural gas, diamonds, manganese, iron ore, cobalt, bauxite, copper, gold, nickel, tantalum, silica sand, clay, cocoa beans, coffee, palm oil, hydropower	51.0	0.462	5.1	ND/Poor	Parliament/ Constitution/ Branches of Government	Republic/ multi-party/ suffrage	0.462	Army, Navy, Air Force [19]

The Country Capacity ID 89

Djibouti	0.89	potential geothermal power, gold, clay, granite, limestone, marble, salt, diatomite, gypsum, pumice, petroleum	68.0	0.470	NA	ND/Poor	Parliament/ Constitution/ Branches of Government	Republic/ multi-party/ suffrage	0.152	Army (includes Navy and Air Force) [38]
Egypt	91.51	petroleum, natural gas, iron ore, phosphates, manganese, limestone, gypsum, talc, asbestos, lead, rare earth elements, zinc	72.0	0.690	1.7	ND/Poor	Parliament/ Constitution/ Branches of Government	Republic/ multi-party/ suffrage	4.783	Army, Navy, Air Force [52]
Equatorial Guinea	0.85	petroleum, natural gas, timber, gold, bauxite, diamonds, tantalum, sand and gravel, clay	86.0	0.587	15.2	ND/Poor	Parliament/ Constitution/ Branches of Government	Republic/ multi-party/ suffrage	1.933	National Guard (with Navy & Air wing) [100]
Eritrea	5.23	gold, potash, zinc, copper, salt, possibly oil and natural gas, fish	59.0	0.391	NA	ND/Poor	Parliament/ Constitution & Branches of Government are in transition	Transitional Government/ single-party/ suffrage	0.046	Ground Forces, Navy, Air Force [1]
Ethiopia	99.39	small reserves of gold, platinum, copper, potash, natural gas, hydropower	43.0	0.442	28.2	ND/Poor	Parliament Constitution Branches of Government	Republic/ multi-party/ suffrage	1.200	Ground Forces, Air Force [4]

(Continued)

Exhibit 5.2 Continued

Country	Population (million)	Resources of Commercial Importance	Literacy Percentage (age 15 and over can read & write)	Human Development Index (HDI)	Capital Formation (investment) % Growth 2014	Infra-structure*	Institutions – Rule of Law	Citizen Participation	FDI 2014 ($ billion)	Military/[Military Spending Budget Per Capita – through 2015 ($US)]
Gabon	1.73	petroleum, natural gas, diamonds, niobium, manganese, uranium, gold, timber, iron ore, hydropower	64.0	0.684	−0.8	ND/Poor	Parliament/Constitution/Branches of Government	Republic/multi-party/suffrage	0.973	Army, Navy, Air Force [72]
Gambia, The	1.99	fish, clay, silica sand, titanium (rutile and ilmenite), tin, zircon	40.0	0.441	0.7	ND/Poor	Parliament/Constitution/Branches of Government	Republic/multi-party/suffrage	0.028	Army, Navy [2]
Ghana	27.40	gold, timber, industrial diamonds, bauxite, manganese, fish, rubber, hydropower, petroleum, silver, salt, limestone	58.0	0.579	3.3	ND/Poor	Parliament/Constitution/Branches of Government	Republic/Constitutional Democracy/multi-party/suffrage	3.364	Army, Navy, Air Force [4]
Guinea	12.61	bauxite, iron ore, diamonds, gold, uranium, hydropower, fish, salt	30.0	0.411	0.8	ND/Poor	Parliament/Constitution/Branches of Government	Republic/multi-party/suffrage	0.567	Army, Navy, Air Force [11]

Guinea-Bissau	1.85	fish, timber, phosphates, bauxite, clay, granite, limestone, unexploited deposits of petroleum	43.0	0.420	NA	ND/Poor	Parliament/ Constitution & Branches of Government are in transition	Republic/ multi-party/ suffrage	0.023	Army, Navy, Air Force [6]
Kenya	46.05	limestone, soda ash, salt, gemstones, fluorspar, zinc, diatomite, gypsum, wildlife, hydropower	86.0	0.548	7.8	ND/Poor	Parliament Constitution Branches of Government	Republic/ multi-party/ suffrage	0.945	Army, Navy, Air Force [13]
Lesotho	2.14	water, agricultural and grazing land, diamonds, sand, clay, building stone	8.5	0.497	NA	ND/Poor	Parliament/ Constitution/ Branches of Government	Constitutional monarchy/ multi-party/ suffrage	0.048	Army (includes Air wing [24]
Liberia	4.51	iron ore, timber, diamonds, gold, hydropower	58.0	0.430	0.3	ND/Poor	Parliament/ Constitution/ Branches of Government	Republic/ multi-party/ suffrage	0.364	Army, Navy, Air Force [2]
Libya	6.28	petroleum, natural gas, gypsum	83.0	0.724	NA	ND/Poor	Parliament/ Constitutional Declaration/ Branches of Government	Transitional Government/ multi-party/ suffrage	0.051	Army, Navy, Air Force [477]
Madagascar	24.24	graphite, chromite, coal, bauxite, rare earth elements, salt, quartz, tar sands, semiprecious stones, mica, fish, hydropower	69.0	0.510	2.2	ND/Poor	Parliament/ Constitution/ Branches of Government	Republic/ multi-party/ suffrage	0.352	Army, Navy, Air Force [2]

(*Continued*)

Exhibit 5.2 Continued

Country	Population (million)	Resources of Commercial Importance	Literacy Percentage (age 15 and over can read & write)	Human Development Index (HDI)	Capital Formation (investment) % Growth 2014	Infra-structure*	Institutions – Rule of Law	Citizen Participation	FDI 2014 ($ billion)	Military/[Military Spending Budget Per Capita – through 2015 ($US)]
Malawi	17.22	limestone, arable land, hydropower, unexploited deposits of uranium, coal, and bauxite	63.0	0.445	7.5	ND/Poor	Parliament/ Constitution/ Branches of Government	Republic/ multi-party/ suffrage	0.717	Army, (includes Air and Navy) [2]
Mali	17.60	gold, phosphates, kaolin, salt, limestone, uranium, gypsum, granite, hydropower; bauxite, iron ore, manganese, tin, and copper deposits are known but not exploited	47.0	0.419	31.3	ND/Poor	Parliament/ Constitution/ Branches of Government	Republic/ multi-party/ suffrage	0.200	Army, Air Force [8]
Mauritania	4.07	iron ore, gypsum, copper, phosphates, diamonds, gold, oil, fish	51.2	0.506	−2.6	MD/needs upgrading	Parliament/ Constitution/ Branches of Government	Democratic Republic/ multi-party/ suffrage	0.503	Army, Navy, Air Force [19]
Mauritius	1.28	arable land, fish	84.4	0.777	−4.6	ND/Needs Upgrading	Parliament/ Constitution/ Branches of Government	Parliamentary Democracy/ multi-party/ suffrage	0.390	No regular military forces [16]
Morocco	34.38	phosphates, iron ore, manganese, lead, zinc, fish, salt	52.3	0.628	−4	ND/Poor	Parliament/ Constitution/ Branches of Government	Constitutional Monarchy/ multi-party/ suffrage	3.583	Army, Navy, Air Force [98]

Mozambique	27.98	coal, titanium, natural gas, hydropower, tantalum, graphite	47.8	0.446	16.5	ND/Poor	Parliament/Constitution/Branches of Government	Republic/multi-party/suffrage	5.000	Army, Navy, Air Force [3]
Namibia	2.46	diamonds, copper, uranium, gold, silver, lead, tin, lithium, cadmium, tungsten, zinc, salt, hydropower, fish, suspected deposits of oil, coal, and iron ore	85.0	0.628	47.5	ND/Poor	Parliament/Constitution/Branches of Government	Republic/multi-party/suffrage	0.494	Army, Navy, Air Wing [125]
Niger	19.90	uranium, coal, iron ore, tin, phosphates, gold, molybdenum, gypsum, salt, petroleum	28.7	0.348	19.6	ND/Poor	Parliament/Constitution/Branches of Government	Republic/multi-party/suffrage	0.770	Army, Navy, Air Force [4]
Nigeria	182.20	natural gas, petroleum, tin, iron ore, coal, limestone, niobium, lead, zinc, arable land	68.0	0.514	13	ND/Poor	Parliament/Constitution/Branches of Government	Republic/multi-party/suffrage	4.657	Army, Navy, Air Force [13]
Rwanda	11.61	cassiterite, wolframite, tantalite, tantalum, beryl, gold, colton, tungsten, niobotantalite, semi-precious stones	70.4	0.483	9.6	ND/Poor	Parliament/Constitution/Branches of Government	Republic/multi-party/suffrage	0.293	Army, Air Force [7]

(*Continued*)

Exhibit 5.2 Continued

Country	Population (million)	Resources of Commercial Importance	Literacy Percentage (age 15 and over can read & write)	Human Development Index (HDI)	Capital Formation (investment) % Growth 2014	Infra-structure*	Institutions – Rule of Law	Citizen Participation	FDI 2014 ($ billion)	Military/[Military Spending Budget Per Capita – through 2015 ($US)]
Sao Tome and Principe	0.19	fish, hydropower	84.9	0.555	NA	ND/Poor	Parliament Constitution Branches of Government	Republic/ multi-party/ suffrage	0.028	Army, Coast Guard [N/A]
Senegal	15.13	fish, phosphates, iron ore	39.3	0.466	2.3	ND/Poor	Parliament/ Constitution/ Branches of Government	Republic/ multi-party/ suffrage	0.344	Army, Navy, Air Force [11]
Seychelles	0.096	fish, copra, cinnamon trees	91.8	0.772	NA	ND/Poor	Parliament/ Constitution/ Branches of Government	Republic/ multi-party/ suffrage	0.109	Army, Coast Guard (includes Naval & Air Wing) [114]
Sierra Leone	6.45	diamonds, titanium ore, bauxite, iron ore, gold, chromite	35.1	0.413	NA	ND/Poor	Parliament/ Constitution/ Branches of Government	Constitutional Democracy/ multi-party/ suffrage	0.691	Army (includes Navy and Air Wing) [6]
Somalia	10.79	uranium and largely unexploited reserves of iron ore, tin, gypsum, bauxite, copper, salt, natural gas, likely oil reserves	37.8	NA	NA	ND/Poor	Parliament/ Constitution/ Branches of Government	Parliament/ multi-party/ suffrage	0.106	National Armed Forces [5]

Country							
South Africa	54.49	gold, chromium, antimony, coal, iron ore, manganese, nickel, phosphates, tin, rare earth elements, uranium, gem diamonds, platinum, copper, vanadium, salt, natural gas	86.4	0.666	1.4	ND/Needs up-grading and development particularly in low-income areas	Parliament/ Constitution/ Branches of Government	Republic/ multi-party/ suffrage	5.741	Army, Navy, Air Force [85]
South Sudan	12.34	hydropower, fertile agricultural land, gold, diamonds, petroleum, hardwoods, limestone, iron ore, copper, chromium ore, zinc, tungsten, mica, silver	27.0	0.467	−2.9	ND/Poor	Parliament/ Constitution & Branches of Government are in transition	Presidential Republic/ multi-party/ suffrage	(0.700)	Army [44]
Sudan	40.24	petroleum; small reserves of iron ore, copper, chromium ore, zinc, tungsten, mica, silver, gold, hydropower	61.1	0.479	−2.2	ND/Poor	Parliament Constitution Branches of Government	Government of National Unity/ multi-party/ suffrage	1.251	Army, Navy, Air Force [61]
Swaziland	1.29	asbestos, coal, clay, cassiterite, hydropower, forests, small gold and diamond deposits, quarry stone, and talc	81.6	0.531	5	ND/Poor	Parliament/ Constitution/ Branches of Government	Monarchy/ Political Associations/ suffrage	0.027	Army, (includes Air wing) [68]

(Continued)

The Country Capacity ID   95

96 The Country Capacity ID

Exhibit 5.2 Continued

Country	Population (million)	Resources of Commercial Importance	Literacy Percentage (age 15 and over can read & write)	Human Development Index (HDI)	Capital Formation (investment) % Growth 2014	Infra-structure*	Institutions – Rule of Law	Citizen Participation	FDI 2014 ($ billion)	Military/Military Spending Budget Per Capita – through 2015 ($US)]
Tanzania	53.47	hydropower, tin, phosphates, iron ore, coal, diamonds, gemstones, gold, natural gas, nickel	69.4	0.521	1.9	ND/Poor	Parliament/ Constitution/ Branches of Government	Republic/ multi-party/ suffrage	2.045	Army, Navy, Air Force [4]
Togo	7.31	phosphates, limestone, marble, arable land	60.9	0.484	2	ND/Poor	Parliament/ Constitution/ Branches of Government	Transitional Republic/ multi-party/ suffrage	0.292	Army, Navy, Air Force [4]
Tunisia	11.25	petroleum, phosphates, iron ore, lead, zinc, salt	74.3	0.721	NA	ND/Poor	Parliament/ Constitution/ Branches of Government	Republic/ multi-party/ suffrage	1.001	Army, Navy, Air Force [49]
Uganda	39.03	copper, cobalt, hydropower, limestone, salt, arable land, gold	66.8	0.483	3.4	ND/Poor	Parliament/ Constitution/ Branches of Government	Republic/ multi-party/ suffrage	1.147	Army, Air Force [7]
Western Sahara	0.57	phosphates, iron ore	NA	NA	NA	ND/Poor	Parliament Constitution Branches of Government	Disputed Territory	NA	Army, Navy [N/A]
Zambia	16.21	copper, cobalt, zinc, lead, coal, emeralds, gold, silver, uranium, hydropower	80.6	0.586	NA	ND/Poor	Parliament/ Constitution/ Branches of Government	Republic/ multi-party/ suffrage	0.508	Army, Navy, Air Force [15]

| Zimbabwe | 15.60 | coal, cobalt, zinc, lead, coal, emeralds, gold, silver, uranium, hydropower | 86.5 | 0.509 | 12.7 | ND/Poor | Parliament/ Constitution/ Branches of Government | Parliamentary Democracy/ multi-party/ suffrage | 0.545 | Army, Navy, Air Force [15] |

Source: World Factbook. Washington DC: Central Intelligence Agency (USA). 2015 (CIA World Factbook—the Best Country factbook available online); GDP(PPP): The World Bank. http://data.worldbank.org; Population (approximate): World Population 2015. Department of Economic and Social Affairs. Population Division. United Nations. 2015; Resources of Commercial Importance: The World Factbook 2013–14. Washington, DC: Central Intelligence Agency, 2013. https: www.cia.gov/library/publications/resources/the-world-factbook/index.html; Literacy (approximate); Human Development Index (HDI): 2014 data: Development Report 2015. United Nations Development Program. http://hdr. undp.org 2014 data; http://data.worldbank.org; Capital Formation growth rates: The World Bank. http://data.worldbank.org; *Infrastructure: ND = Needs Development; MD = Moderately Developed; Infrastructure: Ordinarily, infrastructure is included as capital formation. It is shown separately here for special emphasis; FDI: 2014 data: http://data.worldbank.org; Military Spending Budget data through (2015): www.globalfirepower.com/defense-spending-budget.asp; Defense spending budget data indicates funds allotted to the maintenance and strengthening of a standing military (see globalfirepower); Military spending per capita = military spending budget divided by population.

98 The Country Capacity ID

direction in prescribing a means of remedying Africa's capacity innovation dilemma.

5.1 Population

As in Exhibit 5.2, population size offers a good starting point in assessing capacity ID as it suggests how likely a country may be able to accommodate enterprise of any efficient scale. In this context, African countries for the most part have relatively small populations; smallness, as pointed out in Chapter 3, often has the unique problem of critical mass – too few people, too few resources, too few skills, too little capital, etc. Owing to the population issue, African countries typically lack sizeable domestic markets which in part preclude critical mass and economies of scale and, importantly, economic takeoff. Too few people impede efficient scale enterprise and constrain a country's ability to independently establish market forces that ultimately give rise to self-sustaining markets. Using the G8 – the largest industrialized democracies – as a gauge, the country with the smallest population, Canada at about 36 million today, is larger than the majority of African countries. In fact, the populations of 27 African countries are below 12 million and 19 are below 5 million. Although African countries are well below the capacity of G8 countries, such comparison makes a clear statement about deficiencies in Africa's structural makeup. Too few people amplify matters of critical mass and economies of scale, thus giving population size important implication for dominant or deficient capacity attributes.

5.2 Resources of Commercial Importance

Unlike the population issue, African countries are particularly endowed with natural resources of commercial importance. Exhibit 5.2 reveals the vastness of these resources and other assets possessed by the full gamut of African countries. The holdings of several are particularly noteworthy. Algeria has the world's eighth-largest reserves of natural gas and is the fourth-largest gas exporter; it ranks 14th in oil reserves. Angola is the second largest oil producer and exports more petroleum than any other nation in sub-Saharan Africa and is one of the continent's richest countries. It is the third largest producer of diamonds in Africa and has only explored 40% of its diamond-rich territory. Gabon is on a pace to become the world's leading manganese producer. Natural gas, timber, petroleum, iron, gold, diamonds, niobium, uranium and hydropower are all part of the main resources found abundantly in Gabon. Guinea possesses almost half of the world's bauxite reserves and is the second-largest bauxite producer.[4] Libya ranks 7th in the world in oil reserves and is the fourth largest African country based on oil output.[5] Mauritania has extensive deposits of iron ore; its coastal waters are among the richest fishing areas in the world. Namibia is Africa's fourth-largest exporter of nonfuel minerals and the world's fifth-largest producer of uranium. Its rich alluvial diamond deposits make the country a primary source for gem-quality diamonds. Niger is home to some of the world's largest uranium deposits. Nigeria is Africa's largest oil exporter and the world's 10th largest oil producer. Sierra Leone is a primary source for gem-quality diamonds owing to its alluvial diamond mining operation.[6] South Africa's platinum and manganese reserves are the largest in the world. It is a leading producer of chromite ore with more than 70% of the world's chromite reserves. The country produces more than 10% of the world's gold and has about 50% of the world's gold reserves. It has more than 80% of the world's platinum reserves and produces about 23% of the world's vanadium.[7] And, Togo is the world's fourth-largest producer of phosphate. These impressive statistics underscore the narrative that Africa is the richest continent in natural resources on the globe[8] with proven aggregate reserves of more than 40 percent of the global chromium, 90 percent of the cobalt, 50 percent of the diamonds, more than 50 percent of the gold, 65 percent of the manganese, roughly 8 percent of the oil and gas and most of the platinum and palladium reserves. Africa hosts an abundance of other natural resources including copper, uranium, iron, silver, lead, bauxite, phosphates, timber, 40 percent of the world's potential hydro-electric power, around 60 percent of the world's uncultivated arable land and much more.[9]

5.3 Literacy

Another important core capacity attribute is the literacy of a country's population. A higher level of literacy enables easier skills development thus making possible more efficient technology transfer. UNESCO asserts that "literacy is

---

[4] World Factbook, Washington, DC: Central Intelligence Agency (USA), 2015 (CIA World Factbook – The Best Country factbook available online, www.ciaworldfactbook.us/).
[5] See also http://www.azomining.com/Article.aspx?ArticleID=198.
[6] See World Factbook, ibid.
[7] See http://buzzsouthafrica.com/south-africa-natural-resources/.
[8] Walter E. Williams, "Africa: A Tragic Continent. But Why?" *The New American*, October 29, 2014.
[9] See https://www.credit-suisse.com/us/en/articles/articles/news-and-expertise/2013/05/en/will-africa-s-natural-resources-lead-to-prosperity.html. See also World Factbook, ibid.

a fundamental human right and the foundation for lifelong learning. It is fully essential to social and human development in its ability to transform lives. For individuals, families, and societies alike, it is an instrument of empowerment to improve one's health, one's income, and one's relationship with the world. The uses of literacy for the exchange of knowledge are constantly evolving, along with advances in technology. From the Internet to text messaging, the ever-wider availability of communication makes for greater social and political participation. A literate community is a dynamic community, one that exchanges ideas and engages in debate."[10]

Literacy is of particular concern for African countries as it will in large part influence the pace of capacity innovation and ultimately play an important role in determining how quickly Africa may be able to close the development gap with other regions of the world. The UNESCO Institute for Statistics finds the global average literacy rate based on people age 15 and over who can read and write to be around 86%.[11] This is significantly higher than Africa's average literacy rate of roughly 62%. Exhibit 5.2 provides a country-by-country literacy exposé for Africa. While a number of countries fair well by comparison to global performance – Botswana, Cape Verde, Republic of the Congo, Equatorial Guinea, Kenya, Lesotho, Libya, Namibia, Sao Tome and Principe, Seychelles, South Africa, Swaziland, Zambia and Zimbabwe – the rates for many other African countries underscore the continent's comparatively dismal performance. Several are of particular concern – Benin, Burkina Faso, Guinea, Niger, Senegal, Sierra Leone, Somalia and South Sudan. Not surprisingly, these countries are among the lowest ranking countries based on a world comparison of GDP per capita (PPP) of 230 countries. Indicating the relative economic capacity, so to speak, of country inhabitants to afford "a basket of goods," assuming that education (used here as a surrogate for literacy) is part of the basket of goods, Exhibit 5.3 informs that inhabitants in the lowly ranking African countries simply cannot afford the price of literacy. Put another way, these countries are literally poverty stricken and as such, literacy is a luxury good that to a large extent is beyond their means. The literacy problem is exacerbated by difficulties in sustaining compulsory education in these countries. Although compulsory education is the norm, as conveyed by Exhibit 5.3, resource constraints (shortage of schools, teachers and school supplies) and low attendance – particularly in rural

---

[10] See UNESCO, Literacy, www.unesco.org.
[11] See UNESCO Institute for Statistics, September, 2015, www.uis.unesco.org/literacy.

Exhibit 5.3 Lowest Literacy African Countries GDP Per Capita (ppp) World Ranking.

Rank	Country	GDP per Capita (PPP)* (2015–U.S. Dollars)	Compulsory Education
197	Senegal	2,500	yes/limited research/not fully enforced
204	South Sudan	2,000	yes/limited resources, low attendance
206	Benin	2,000	yes/limited resources, low attendance (between ages 6 and 11)
208	Burkina-Faso	1,800	yes/limited resources, low attendance
215	Sierra Leone	1,600	yes/limited resources, low attendance
220	Guinea	1,300	yes/(between ages 7 and 13) limited resources, low attendance
223	Niger	1,000	yes/limited resoures, low attendance
229	Somalia	400	yes/limited resources, low attendance

Source: The World Factbook, Washington, DC: Central Intelligence Agency. The data are 2015 estimates; Somalia 2014 estimate.

*GDP per capita (ppp) compares GDP on a purchasing power parity basis divided by population. PPPS are the rates of currency conversion that equalize the purchasing power of different currencies by eliminating the differences in price levels between countries.

areas,[12] pose rather insurmountable challenges in many cases. Ultimately, poorly educated or uneducated children become educationally deficient or illiterate adults, hence the comparatively low average literacy rate for Africans age 15 and over.

5.4 Human Development Index

Country capacity ID is further illuminated by improvements in human well-being as conveyed by the human development index (HDI). The HDI is a summary measure of human development focusing on standard of living, literacy and life expectancy. With an upper limit of 1.0, HDI ranks countries in four classifications; very high, high, medium and low human development. Typically, African countries predominantly populate the low human development category. In 2014, for example, the HDI value in the low human development category ranged from .348 (Niger) to .548 (Kenya); 36 of 44 countries in this category were African.[13] As shown in Exhibit 5.2, the highest scoring African country is Mauritius (.777) which places it in the high human

---

[12]Children often must assist with farm chores or engage in other income earning activities to support the family.

[13]See Human Development Report 2015, United Nations Development Programme, pp. 212–215, http://hdr.undp.org.

development category in the human development index along with only four other African countries, Seychelles (.772), Algeria (.736), Libya (.724) and Tunisia (.721). No African country scored in the very high classification. For the most part, Africa has much catching up in the area of human well-being.

5.5 Capital Formation

Capital formation as a core capacity attribute is particularly important owing to its direct connection to capacity innovation and the development process. Recent empirical studies conducted in Africa establish the critical linkage between capital formation and the pace of development.[14] Further, private capital formation has a stronger, more favorable effect on development than government capital formation because private capital formation is more efficient and less closely associated with special interest impulses.[15]

Capital formation in this narrative connotes capital accumulation or increase in capital stock – a process of deployment of physical capital as investment in fixed assets (buildings and machinery or plant and equipment). Meier confides that "Throughout the history of economic thought, the accumulation of real physical capital stock, has been viewed as permitting more roundabout methods of production and greater productivity, thereby

---

[14] See I.M. Shuaib and Dania Evelyn Ndidi, "Capital Formation: Impact on the Economic Development of Nigeria 1960–2013," *European Journal of Business, Economics and Accountancy*, Vol. 3, No. 3, 2015, p. 2. For a comprehensive literature review on this matter, see Eric Akobeng, "Gross Capital Formation, Institutions and Poverty in Sub-Saharan Africa," *Journal of Economic Policy Reform*, January, 25, 2016, pages 1–29, http://www.tandfonline.com/doi/full. See also Cesar Calderon, Infrastructure and Growth in Africa, *Policy Research Working Paper 4914*, The World Bank, Africa Region, African Sustainable Development Office, April 2009; M.S. Khan and C.M. Reinhart, "Private Investment and Economic Growth in Developing Countries," *World Development*, 18 (1), 1990, pp. 19–27;
D. Ghura and Hadji Michael T., Growth in Sub-Saharan Africa, *Staff Papers*, International Monetary Fund, 43, September, 1996; D. Ben-David, "Convergence Clubs and Subsistence Economies," *Journal of Development Economics*, 55, pp. 155–171 1998; P. Collier and J. W. Gunning, " Explaining African Economic Performance," *Journal of Economic Literature*, 37, March,1999, pp.64–111; E. Herandez-Cata, "Raising Growth and Investment in Sub-Saharan Africa: What Can be Done?" *Policy Discussion Paper: PDP/00/4*, Washington, D.C.: International Monetary Fund, 2000; and L. Ndikumana, "Financial Determinants of Domestic Investment in Sub-Saharan Africa," *World Development*, 28(2), 2000, pp. 381–400.

[15] See Shuaib and Ndidi, ibid., pp. 3–4.

providing an additional future stream of income to society"[16] purposely enabling capacity innovation. A nation's "rate of progress is proportional to its rate of investment: When we compare, therefore, the state of a nation at two different periods, and find, that the annual produce of its land and labor is evidently greater at the latter than the former, that its lands are better cultivated, its manufactures more extensive, we may be assured that its capital must have increased during the interval between those two periods"[17] and that capacity innovation must have occurred or improved. "Over the long run, rapid economic growth does not take place without large investment in fixed capital. Many theoretical arguments and empirical studies conclude that the rate of capital formation determines the rate of a country's economic growth."[18] It is common to attribute as much as 50 percent of the increase in GDP to capital formation.[19]

Exhibit 5.2 attempts a broad reflection of the state of capital formation in African countries. The general impression is that investment in the capital stock is quite small and even negative in a number of cases – Comoros, Equatorial Guinea, Gabon, Mauritania, Mauritius, Morocco, South Sudan and Sudan. As Seth observes, not only is the existing stock of capital very small in African countries, but the current rate of capital formation is also very low.[20] Exhibit 5.4 allows a more careful examination of this matter.

Exhibit 5.4 Africa and G8 Relationship of Capital Formation to Gross Domestic Product (2014 – Average).

Country Grouping	Capital Formation (Annual Growth Rate)	Capital Formation (% of GDP)	*GDP per Capita (PPP) (Current Int'l $)	GDP (PPP) Millions (Current Int'l $)
Africa	7.22	25.38	5859	104,140
G8	.51	20.0	40,090	4,768,664

Source: Raw data provided by The World Bank, http://data.worldbank.org/indicator/RIY.GDP.PCAP.PP.CD/countries.
*GDP per capita based on purchasing power parity (PPP). PPP GDP is gross domestic product converted to international dollars using purchasing power parity rates. An international dollar has the same purchasing power over GDP as the U.S. dollar has in the United States.

---

[16] Meier, op. cit., p. 163.

[17] Ibid.

[18] Eric Akobeng, "Gross Capital Formation, Institutions and Poverty in Sub-Saharan Africa," *Journal of Economic Policy Reform*, January 25, 2016, p. 9.

[19] Meier, op. cit., p. 164.

[20] Tushar Seth, *Reasons for Low Capital Formation in Under-Developed Countries*, http://www.economicsdiscussion.net/articles/reasons-for-low-capital-formation-in-under-developed-countries/1537.

Interestingly, capital formation in Africa has been rather impressive by comparison to the G8. Increase in capital stock in African countries has out-paced G8 countries more than seven to one; the average annual growth rate in Africa stands at 7.22% and the G8 at a mere ½ percent. Africa has also performed comparatively well when considering capital formation as a share of GDP; here accounting for about ¼ of Africa's GDP but ⅕ of the G8. But there are two sides to this assessment. African capital formation performance relative to the G8 may be somewhat deceiving as it perhaps masks other important factors that should be addressed alongside capital stock. Of particular note is Africa's comparative purchasing power parity GDP performance. Despite the state of capital formation in Africa and its apparent impact on gross domestic product, the monetary value of the finished goods and services produced in African countries based on purchasing power parity does not measure up to the G8, not even reasonably so.[21] The purchasing power parity GDP per capita and purchasing power parity GDP as shown in Exhibit 5.4 reflect vastly different comparative outcomes. Average GDP per capita (PPP) for the G8 is nearly seven times greater than Africa's performance and average GDP (PPP) for the G8 is nearly 46 times greater. The results beg the question: why is Africa's rather superior performance in capital formation relative to the G8 not reflected in relative gross domestic product measures? The explanation is perhaps twofold. First, Africa's rate of increase in capital stock is based on rather small economies which essentially distort the full value and meaning of the rate itself. For example, Namibia with a population of about 2.5 million has a 2014 capital formation growth rate above 47 percent but its economy only measures at about $24 billion GDP (PPP). By contrast, Canada with a population of about 36 million – the smallest in the G8, only grew at –0.9 percent (a loss) capital formation-wise but the size of its economy approximates $1.6 trillion GDP (PPP). This gives rise to the second explanation. African countries have not reached economic take-off, as in the G8, whereby there is large investment in fixed capital and sustained growth across the full range of economic activity that becomes a normal condition enabled by advantages of critical mass and

---

[21]Purchasing power parity essentially approximates the law of one price. "It asserts that identical goods should cost the same in all nations, assuming that it is costless to ship goods between nations and there are no barriers to trade (such as tariffs). Before the costs of goods in different nations can be compared, prices must first be converted into a common currency. Once converted at the going market-exchange rate, the prices of identical goods from any two nations should be identical." See Robert J. Carbaugh, *International Economics*, Eighth Edition, Cincinnati, Ohio: South-Western, 2002, p. 403.

economies of scale, as discussed in Chapter 3. Further, capacity innovation in the G8 is regenerative as a result of economic take-off; such is not the case in African countries. It is perhaps unlikely that any African country standalone would be able to approximate the performance of a G8 member country owing to the problems of critical mass and economies of scale, among others; as Exhibit 5.2 reveals, 45 of the 55 African countries (Western Sahara disputed) have populations smaller than Canada and none has a larger GDP (PPP). Looking ahead, Chapter 7 proposes a fix for the problems of critical mass and economies of scale.

5.6 Infrastructure

The World Bank observes infrastructure as capital formation or fixed investment, thereby using capital formation in its more holistic delineation: land improvements (fences, ditches, drains, and so on); plant, machinery, and equipment purchases; and the construction of roads, railways, and the like, including schools, offices, hospitals, private residential dwellings, and commercial and industrial buildings.[22] It is used separately here merely to emphasize its importance to capacity innovation. As with capital formation, infrastructure is directly connected to capacity innovation and Exhibit 5.2 attempts a broad reflection of the condition of African infrastructure. Under the general impression that there exists a strong need for infrastructure development in African countries and that existing infrastructure is in a poor state, needs development/poor is used to so indicate. These views are supported by empirical literature on capacity innovation that identifies inadequate infrastructure provision as one of the main problems in Africa.[23] Further, UNCTAD asserts that Africa must lift the main binding constraints to investment, among them – the poor state of infrastructure.[24] And, research by the World Bank indicates that the impact of infrastructure development on capacity innovation in African countries is decidedly positive and that "African countries are likely to gain more from larger stocks of infrastructure than from enhancements in the quality of existing infrastructure."[25] The Economic Commission for Africa shares these views noting that "the ability

---

[22]The World Bank, http://data.worldbank.org/indicator/NE.GDI.FTOT.ZS.
[23]Cesar Calderon, op. cit., p. iv.
[24]*The Economic Development in Africa Report 2014*, UNCTAD/ALDC/AFRICA/2014, New York: United Nations (UNCTAD), 2014.
[25]Cesar Calderon, op. cit., pp. i and 19.

of African countries to establish a competitive industrial sector and promote greater industrial linkages has been hindered by poor infrastructure (energy, transport, communication, etc.), which has resulted in high production and transaction costs. It is therefore imperative to drum up massive investments in infrastructure, including energy. The development of infrastructure, therefore, has to be made a priority at national, regional and continental levels. Regional cooperation in the development of infrastructure would lower transaction costs, enhance regional markets and make production and exports more competitive."[26] In this regard, Yang and Gupta assert that Africa must concentrate more investment in infrastructure development if it is to significantly improve regional trade and that the quality of infrastructure or lack thereof raises transport costs by 12 percent and reduces trade volume by 28 percent.[27] There is therefore a tremendous need to develop and upgrade Africa's poor infrastructure much of which quite frankly was built during the immediate independence era.[28]

Identifying a country-by-country general condition of African infrastructure as 'needs development/poor,' Exhibit 5.2 recognizes only two African countries having infrastructure falling outside this description – Mauritius and South Africa. Interestingly, the World Bank also views these countries as having any credible infrastructure capacity suggesting that "if all African countries were to catch up with Mauritius in infrastructure, per capita economic growth in the region could increase by 2.2 percentage points" and that Africa "will require a significant expansion of water storage capacity from current levels of 200 cubic meters per capita to levels of at least 750 cubic meters per capita, a level currently found only in South Africa."[29] Africa's largest infrastructure deficit is in the power sector which delivers only a

---

[26] United Nations Economic Commission for Africa, http://www.uneca.org/.

[27] See Benjamin F. Bobo and Hermann Sintim-Aboagye, eds., *Neo-Liberalism, Interventionism and the Developmental State: Implementing the New Partnership for Africa's Development*, Trenton, New Jersey: Africa World Press, 2012, p. 214 and Yongzheng Yang and Sanjeev Gupta, "Regional Trade Arrangements in Africa: Past Performance and the Way Forward," African Development Bank, Blackwell Publishing Ltd., pp. 399–431.

[28] Karl Botchway and Jamee Moudud, Neo-Liberalism and the Developmental State: Consideration for the New Partnership for Africa's Development," Chapter 2, in Benjamin F. Bobo and Hermann Sintim-Aboagye, eds., *Neo-Liberalism, Interventionism and the Developmental State: Implementing the New Partnership for Africa's Development*, Trenton, New Jersey: Africa World Press, 2012, p. 17.

[29] The World Bank, *Fact Sheet: Infrastructure in Sub-Saharan Africa*, http://go.worldbank.org/SWDECPM5S0.

fraction of the service found elsewhere in the developing world.[30] "The 48 countries of Sub-Saharan Africa (with a combined population of 800 million) generate roughly the same amount of power as Spain (with a population of 45 million)."[31] Clearly, the state of African infrastructure is problematic and poses tremendous challenges to development; as a core capacity attribute, it threatens Africa's ability to navigate the ladder of comparative advantage.

5.7 Institutions – Rule of Law

The importance of institutions and the rule of law in establishing a system of strong core capacity attributes cannot be overstated. As conveyed in Chapter 2, requisite to the governing process is an institutional mapping that prescribes as well as implements an organization of governmental decision-making units – branches of government, complete with a government philosophy or set of guiding principles – a constitution, procedures for implementing the guiding principles, and rules of law to assure that the guiding principles and procedures are observed and practiced. These are integral to the governing process and can directly assist government in promoting and coordinating policy objectives as they provide frameworks and mechanisms to facilitate desired outcomes.[32] Institutions and the rule of law are not simply important to the governmental process but to the market process as well. As asserted in Chapter 2, occasionally the market mechanism is assisted by institutions as they provide oversight and corrective action to improve market performance.

In the African context, Exhibit 5.2 informs that African countries display an institutional makeup characterized by a constitution, branches of government including a parliament and through these foundational institutional arrangements, implied rule of law and the democratic process. These may be presumed to be essential to effective development and capacity innovation strategy. Botchway and Moudud argue that an effective development strategy can only take place on the basis of national consensus, hence the need for the democratic process.[33] "The efficacy of these institutional arrangements it is believed would lead to what the mainstream literature terms – good

---

[30]The World Bank, ibid.

[31]The World Bank, ibid.

[32]See Benjamin F. Bobo, "Implementing State Interventionist Development: The Role of the Multinational Corporation," ibid., pp. 221–222; and Hilton L. Root, *Small Countries, Big Lessons: Governance and the Rise of East Asia*, op. cit.

[33]Botchway and Moudud, op. cit., p. 33.

governance."[34] Although these institutions alone do not necessarily translate to the democratic process, the presence of an executive branch with an elected head of state, a legislature with a bicameral parliament and a judiciary headed by a supreme court, as typically characteristic of African governments, suggest that the key elements of a democratic institutional structure accompanied by rule of law are in place. Without these, the democratic process is questionable; with these, it is fully possible.

The rule of law is a particularly compelling core capacity attribute as it stands essentially as a nation's supreme modus operandi: the rule of law presides over the functioning of a nation; a nation is governed by law and all citizens including government are subject to it; no one or no entity is above it; the law applies to everyone equally. Absent the rule of law, the supreme power of a nation may be centralized to an individual who may exercise power by personal decree. Even with a rather standard institutional mapping in place, many African leaders are viewed as repressive, overreaching and autocratic. This is supported by the Ibrahim Index of African Governance (IIAG) which measures the quality of governance in African countries based on the rule of law and a host of other factors. The impression gleaned from the IIAG is that the overall governance in Africa has stalled since 2011. With 100 the highest score possible, in 2015 all African countries scored below 80, 27 scored below 50 and the average score was barely above 50.[35] Although African countries in the main have adopted the rule of law, as discernable from Exhibit 5.2, the IIAG suggests that they generally fall well short of observing it.

5.8 Citizen Participation

For the most part, African countries are republics – sovereign nations, states or political orders in which the supreme power is held by the citizens who are entitled to vote for government representatives including presidents responsible to them and power is exercised according to the rule of law; display multi-party arrangements – more than two political parties; and subscribe to suffrage – the right or privilege to vote in political elections. These institutional elements provide the foundational structure for citizen participation as a core capacity attribute. The right of the individual to choose representatives and to bestow power upon them is fundamental to participation in the governance process and promotion of democratic government forms. This is perhaps the

---

[34] Ibid., p. 35.
[35] Mo Ibrahim, 2015 Ibrahim Index of African Governance, http://static.moibrahimfounda tion.org/u/2015.

highest form of citizen participation which in a broader context is a process in which individuals have an opportunity to influence public decisions through input to the democratic decision-making process.[36] In an increasingly evolving and interconnected world, marked by global movement towards democratic institutions, citizen participation offers opportunities to strengthen democracy, accountability and the rule of law.[37]

While the citizen participation apparatus is common among African countries, mobilizing the citizenry is not without its challenges. "Mutual and growing distrust deepens the gap between citizens and their leaders. Persistent repression and the manipulation of laws and public institutions for political purposes regularly stifle perceived opposition. Moreover, for civil society organizations to effectively contribute to improving governance in Africa, their own issues of capacity and resources, as well as competition, internal governance, representation and legitimacy must be addressed. In several countries, the co-existence of local and international non-governmental organizations – with the latter offering better employment conditions – often leads to a local "brain-drain" that deprives national and community-based organizations of much needed capacities and resources. This loss of expertise and undue competition with better resourced international organizations also have the potential to negatively impact local civil society's role as their communities change agents."[38] For citizen participation to stand as a strong core capacity attribute, such challenges must be remedied.

5.9 Foreign Direct Investment (FDI)

Foreign direct investment[39] is a leading core capacity attribute and a key driver of African development. Attracting sustained private investment from external sources is high among Africa's development priorities and interest in

---

[36] Pages.uoregon.edu/rgp/PPPM613/class10.

[37] M. A. Mindzie, "Citizen Participation and the Promotion of Democratic Governance in Africa," *Great Insights Magazine*, Volume 4, Issue 3, April/May, 2015; http://ecdpm.org/great-insights/rising-voices-africa/citizen-participation-promotion-democratic-governance-africa/.

[38] Ibid.

[39] Foreign direct investment is a class of investment that reflects the objective of establishing a long term interest by a direct investor or resident enterprise in a country other than that of the direct investor. FDI encourages the transfer of technology and know-how between countries and allows the host economy to promote its products more widely in the global marketplace. Under the right policy environment, FDI can be an important vehicle for development. See OECD (2016), "Foreign direct investment flows," in *OECD Factbook 2015–2016: Economic, Environmental and Social Statistics*, Paris: OECD Publishing.

capturing the attention of multinational corporations is quite evident. Signaling this interest, the New Partnership for Africa's Development stresses that there is an urgent need to create conditions conducive to private sector investments by foreign investors and seeks to increase private capital flows to Africa as an essential component of a sustainable long-term approach to filling the capital resource need through promotion of foreign direct investment and trade.[40] "FDI is generally a steadfast source of capital that supports macroeconomic stability and allows emerging economies in particular to quickly ramp up their capital stock and knowledge economy."[41]

Foreign investment on the continent reached about $60 billion in 2014, a tidy sum generally speaking but for individual African countries much is left unsaid. While this amount averages to only about $1.1 billion per country, a closer examination as depicted in Exhibit 5.2 reveals that investment in many cases is far below the average. In 26 African countries FDI is around $⅓ billion or less and in none is it greater than $5.75 billion. A G8 comparison offers perspective here. An aggregate FDI for Africa in 2014 of $60 billion across a population of 1.18 billion generates an FDI per capita of about $50 whereas a $540 billion FDI for the G8 covering a population of roughly 883 million yields about $600 FDI per capita. Obviously a G8 comparison is a far reach but it is a rather sobering indication of Africa's economic position by comparison to the largest industrialized group of countries.

While multinationals are not avoiding investment in the continent, it is clear that Africa's persistent lagging in capacity innovation and sustained development suggests that the level of foreign direct investment, hence the attention paid by multinationals to business opportunity on the continent, does not meet its needs. The New Partnership for Africa's Development acknowledges that "many countries lack the necessary policy and regulatory frameworks for private sector-led growth," hence their inability to more significantly attract multinationals.[42] Further, NEPAD recognizes that "capacity [innovation] is a critical aspect of creating conditions for development" and that African countries have "a major role to play in promoting economic growth and development" but "the reality is that many governments lack the capacity to fulfil this role."[43] A more forward looking investment

---

[40]The New Partnership for Africa's Development (NEPAD), October 2001, articles 145, 150, 163 and 166.

[41]"Insuring FDI's Success," *Zurich*, October 3, 2014, p. 1, https://www.zurich.com/en/knowledge/articles/2014/10/insuring-fdi-success.

[42]The New Partnership for Africa's Development, ibid., article 86.

[43]Ibid.

posture by African countries particularly one that shows the collective will of African nations may produce far better results. Chapter 7 takes up this matter.

5.10 Military

The importance of the military as a core capacity attribute relates primarily to the protection of foreign direct investment especially where corporate assets may be isolated and in non-urban areas. Protecting foreign investment, particularly of the direct kind, has a long and storied history. Ronald Brand, in *Fundamentals of International Business Transactions,* brings to fore the nature and complexities of protecting FDI and the international entanglements and trappings of concepts and procedures that define direct capital flows to foreign venues.[44] In so doing, Brand focuses on the efforts of capital exporting countries to personally secure their investments abroad – this of course involving multinational corporations. In important ways, FDI is secured by the international investment law regime framework that encompasses, among others, preferential trade and investment agreements (PTIAs), economic integration agreements (EIAs), international investment agreements (IIAs) and Bilateral Investment Treaties (BITs).[45] All in ways address issues related to FDI access, diffusion, treatment and protection. While such arrangements do much to encourage, normalize and stabilize investment relations, major concerns extend to "governance," particularly "rule of law" and "political stability" especially from the standpoint of the perception of the quality of the law enforcement security apparatus in host countries.[46] Dollar asserts that rule of law and political stability are positively correlated; if political stability is excluded, the coefficient on rule of law becomes larger and more significant. In fact, a gain of one standard deviation on rule of law corresponds to about 40% more FDI.[47]

---

[44] Ronald A. Brand, *Fundamentals of International Business Transactions*, The Hague, The Netherlands: Kluwer Law International, 2000.

[45] *The Role of International Agreements in Attracting Foreign Direct Investment to Developing Countries*, New York: UNCTAD, 2009, p. 77; and *Reforming the International Investment Regime: Need for African Voice*, African Center of International Law Practice, September 30, 2017, pp. 1–2.

[46] David Dollar, *United States-China Two-Way Direct Investment: Opportunities and Challenges*, Brookings Institution, January 2015, p. 4.

[47] These findings are based on a study of U.S. and China outward direct investment. See Dollar, ibid., p. 19.

As African countries rely on the military for law enforcement purposes, however limited, the availability of military resources is a strategic core capacity attribute and a key factor in the attraction of foreign direct investment. Deliberating on the case of South Africa, Montesh and Basdeo argue that the role "the South African National Defense Force [the military] can play in precipitating and sustaining law enforcement in civil society"[48] is often overlooked. They assert that by so doing, criminal activity that may otherwise be obviated prevents entrepreneurs and investors from taking advantage of the opportunities that South Africa offers.[49] Further, Montesh and Basdeo purport that "by making use of the collateral utility available in its primary design"[50] the South African National Defense Force by mandate of the South African Defense Act may be deployed with the South African Police Service to perform a number of functions including "... the preservation of life, health or property."[51]

Thus, an alternative approach to FDI security is proposed in this volume whereby the host government assumes responsibility for protecting foreign assets – including physical and human, and places this responsibility under the military command. (A host government protection scheme is formally addressed in Chapter 8.) This approach is not holistically a new concept and certainly not necessarily intended to be mutually exclusive, i.e., the capital recipient's act of protection does not rule out the capital provider's act of protection. Akin to a gendarmerie, FDI security in principle would be assigned to a military force charged with police duties having the capacity, training and weaponry to especially confront armed groups and more intense violence; in urban areas this would be more within the confines of civilian populations but less so in rural areas where capital diffusion may be more dispersed and isolated than in highly populated environs.[52] The African Union's (AU)

---

[48] Moses Montesh, and Vinesh Basdeo, "The Role of the South African National Defense Force in Policing," *South African Journal of Military Studies*, *40*(1), 2012, p. 71.
[49] Ibid.
[50] Ibid., p. 81.
[51] Ibid., p. 78.
[52] See Giovanni Arcudi and Michael E. Smith, "The European Gendarmerie Force: A Solution in Search of Problems?" *European Security*, 22(1), 2013, pp. 1–20; *French National Gendarmerie,* http://www.fiep.org/member-forces/french-national-gendarmerie *and* http://www.gendarmerie.interieur.gouv.fr; Dirk Peters and Wolfgang Wagner, Between Military Efficiency and Democratic Legitimacy: Mapping *Parliamentary* War Powers in Contemporary Democracies, 1989–2004," *Parliamentary Affairs*, Volume 64, issue 1, September 2010, pages 175–192; http://www.eurogendfor.org; https://en.wikipedia.org/wiki/Gendarmerie; https://en.wikipedia.org/wiki/List_of_gendarmeries; and http://www.independent.co.uk/news/world/europe/who-are-gign-elite-police-force-formed-after-1972-olympics-attack-on-israelis.

protocol on African security offers a clear path to introducing this approach. The AU has established and charged the African Union's Peace and Security Council and the African Standby Force (comprised of military resources from member nations) with the responsibility to " promote peace, security and stability in Africa, in order to guarantee the protection and preservation of life and property, the well-being of the African people and their environment, as well as the creation of conditions conducive to sustainable development" and further to "promote and encourage democratic practices, good governance and the rule of law, protect human rights and fundamental freedoms, respect for the sanctity of human life and international humanitarian law . . ."[53] This structure fits well the intent of this approach to place the burden of foreign direct investment security on the host government, in this instance – NEPAD and the African military establishment, where response time may be more efficient, continental prerogative more assured and an African military force more palatable.

This is viewed as a rational approach as Kohler asserts that "the military may increase security which should lead to higher investment and growth."[54] Employing econometric modeling and data from 27 African countries, Kohler endeavored to determine whether the positive effects on investment of increased security outweigh the resource costs in foregone investment of defense. The results were instructive: a well-equipped and well-trained military gives rise to additional investment through its security enhancing effects; allocating more resources to defense leads to higher gross fixed capital formation; and large armies, especially when ill-equipped and ill-trained, have a negative effect on gross fixed capital formation. These results lead to the impression that multinationals are positively influenced by military presence and concentrate higher investment in countries holding well-resourced militaries. Higher capital concentration leads to greater capital formation in targeted countries which in turn results in higher capital stock;[55]

---

[53] See *Protocol Relating to the Establishment of the Peace and Security Council of the African Union* in Appendix.

[54] Daniel F. Kohler, *The Effects of Defense and Security on Capital Formation in Africa*, Santa Monica: The Rand Corporation, N-2653-USDP, September 1988; and https://www.rand.org/content/dam/rand/pubs/notes/2007/N2653.pdf.

[55] Capital formation, gross capital formation, and gross fixed capital formation are often used synonymously but strictly speaking there are definitional differences and the terms are subject to different interpretations. See Simon Kuznets, "Proportion of Capital Formation to National Product " *The American Economic Review*, Vol. 42, No. 2, Papers and Proceedings of the Sixty-fourth Annual Meeting of the American Economic Association (May, 1952), pp. 507–526; http://data.worldbank.org; and http://lexicon.ft.com.

these outcomes potentially lead to higher output and therefore economic growth over time.[56]

The presence of military units as security for physical and human assets particularly where they are well-resourced and deployed with a mandate to protect foreign direct investment has implications for African FDI policy as Exhibit 5.2 shows that African countries typically have a full military apparatus – Army, Navy and Air Force capabilities. The quality of their equipment and training notwithstanding, the security enhancing effects of these resources may serve Africa well particularly in light of NEPAD's mandate "to increase private capital flows to Africa" and in so doing "the first priority is to address investors' perception of Africa as a "high-risk" continent, especially with regards to security of property rights, regulatory frameworks and markets."[57]

Although the military capacity of African countries, see Exhibit 5.2, may be limited as judged by their aggregate military spending per capita of around $43 annually – by comparison the G7 spends roughly $1630 per capita annually,[58] the African Union's African Standby Force may be strengthened by the collective deployment of military forces not only in the traditional role of defense of the state and its citizenry[59] but protection of foreign direct investment as well. In this regard a cautionary note is important here. It is recognized that the frequent use of armed forces may enhance their influence and control in the polity which may threaten governance and democracy. There have been successes as well as failures. Lu, Miethe and Liang note that having the military assist with community-based activities has been employed successfully in China, El Salvador and the Philippines, among others.[60]

---

[56]Kohler, op. cit., p. 42. Unlike Kohler's study, some researchers have found a negative relationship between defense spending and economic growth. Kohler asserts that few of these studies allowed for differences in threats faced by different countries, hence leading to misspecified models in which a relevant "threat" variable is excluded from the growth equation; threat being guerilla activity, an aggressive neighboring country and/or heavy military expenditure in a neighboring country. Kohler argues that since threats and defense spending are closely and positively related to each other, defense spending tends to act as a proxy for threats in such models, thus explaining the negative coefficients estimated. See Kohler, pp. 42–43.

[57]The New Partnership for Africa's Development, ibid., articles 150 and 151.

[58]Defense spending in the G7 (the US, the UK, France, Germany, Italy, Japan and Canada) stands around $660bn a year, see http://www.theguardian.com/politics/2005/jul/06/g8.military.

[59]Montesh and Basdeo, op. cit., p. 78.

[60]H. Lu, DT Miethe and B. Liang, *China's Drug Practices and Policies*, Oklahoma City: Oklahoma State University, Ashgate Publishers, 2009, in Montesh and Basdeo, ibid., p. 87.

But Pakistan and Bangladesh, for example, have suffered military coups.[61] Recommending involvement of the military as proposed, however, is strategically rational, despite potential drawbacks. Capital diffusion in Africa will assume urban and rural locations, perhaps rural even more so. As such, application of a credible law enforcement regime is critical. Police are more effective in urban areas in a law and order function but the military is more effective in rural environs in defense against external military aggressors – their respective training regimens so attest. Police are trained to "fight crime," protect civilians and maintain order[62] whereas "armed forces are not trained, orientated or equipped for deployment against civilians . . . they are trained to kill, decimate and destroy."[63] Working to their respective strengths with one assisting the other as occasions arise allows more efficient outcomes. A military security regime for foreign direct investment signals multinational corporations that Africa is not only open for business but is prepared to provide necessary protection for corporate assets wherever they may be located on the continent. Military security places the level of commitment by African authorities and the level of confidence by multinational corporations outside the scope of local policing, beyond a national objective, and ultimately within an intra-continental security agenda. As this volume speaks to intra-continental capacity innovation whereby African countries collectively pursue integration of development objectives – via NEPAD administration, employing an intra-continental security regime captures the full essence of intra-continental cooperation.

5.11 Closing Observations

It appears that a particular problem for African countries in redressing their capacity innovation dilemma is the absence of a credible system of mutually-reinforcing development factors – such as the core capacity attributes, to significantly enable capacity innovation. Clearly, Africa in all of its 54 parts (excluding the disputed territory of Western Sahara) shows deficiencies in core capacity attributes, quite significantly so in some cases. Africa's strong suit as is apparent is resources of commercial importance. The continent is a vast depository of mineral wealth and many countries hold significant quantities of

---

[61] Montesh and Basdeo address the use of the military in community-based matters in "The Role of the South African National Defense Force in Policing," ibid., pp. 87–91.

[62] Montesh and Basdeo, ibid., pp. 72, 90.

[63] Ibid.

a fairly wide range of mineral assets. But rather weak country capacity IDs give rise to an incapacity to convert these resources to end user products through domestic means and to develop national markets that support enterprise of any efficient scale. Core capacity attributes across African countries bespeak underlying limitations in population size, people age 15 and over who can read and write, human development, capital formation by comparison to other regions, infrastructure capacity, observation of the rule of law, citizen participation, attracting foreign direct investment, and deployment of the military to protect FDI. These observations offer key insight and direction in prescribing a means of remedying Africa's capacity innovation dilemma.

6

Africa-MNC Strategic Alliance

This volume proposes a direct relationship between Africa and the multinational corporation as a means of prompting efficient capacity innovation across the continent. A bountiful approach to this endeavor is through the lens of a strategic alliance. NEPAD as proxy for the African continent in an alliance with the multinational corporation as proxy for the capacity innovation mechanism augmented by participation of the NGO as agent for the grassroots – as articulated in the cross-fertilizing integrative framework in Chapter 4, provide an operative construct for a wide variety of essential features for African capacity innovation such as: motivation – political will, profit motive and MNC objective;[1] interest alignment – foreign direct investment and new market penetration;[2] longevity – sustained development and long-term investment;[3] and input-output solutions – technology transfer, living standards upgrade, shareholder wealth maximization and value creation.[4]

---

[1] Lori Ann Post, Amber N. W. Raile and Eric D. Raile, "Defining Political Will," *Politics & Policy*, 38(4), August 2010, pp. 653–676; and Medard Gabel and Henry Bruner, *Global Inc. An Atlas of The Multinational Corporation*, New York: The New Press, 2003.

[2] K.R. Harrigan, "Vertical Integration and Corporate Strategy," *Academy of Management Journal*, 28(2), pp. 397–425; Kathryn Rudie Harrigan, "Strategic Alliances as Agents of Competitive Change," 1985, https://www0.gsb.columbia.edu/mygsb/faculty/research/pubfiles/10497/Harrigan%20Strategic%20Alliances.pdf; and Kathryn Rudie Harrigan, "Joint Ventures and Competitive Strategies," *Strategic Management Journal*, Volume 9, Issue 2, March–April, 1988, pp. 141–158.

[3] Benjamin F. Bobo, *Rich-Country, Poor Country: The Multinational as Change Agent*, Westport, CT: Praeger Publishers, 2005.

[4] D.C. Movery, J.E. Oxley and B.S. Silverman, Strategic Alliances and Interfirm Knowledge Transfer," *Strategic Management Journal*, 17, pp. 77–91; B.H. Anand and T. Khanna, "Do Firms Learn to Create Value? The Case of Alliances," *Strategic Management Journal*, 21(3), 2000, pp. 295–315; and J.J. Reuer, "From Hybrids to Hierarchies: Shareholder Wealth Effects of Joint Venture Partner Buyouts," Strategic Management Journal, 22(1), 2001, pp. 27–44.

The proposed NEPAD/MNC/NGO arrangement is, as Sanyal suggests, tantamount to a rarely seen "triple alliance"[5] of government, market institutions and NGOs in which the government creates the policy environment necessary for maximizing the effectiveness of the joint arrangement: government (NEPAD) has the administrative machinery to support large-scale implementation of projects and initiatives; market institutions (multinationals) heighten the sensitivity of capacity innovation to the preferences of consumers and producers, and infuse a sense of "market discipline" in the organization of capacity innovation activities; and NGOs ensure participation of citizens, particularly the grassroots, who are beyond the reach of the government's administrative apparatus. In total, NEPAD reflects the interests of the state, MNCs are driven by the profit motive and NGOs are motivated by community needs; when interacting in a cross-fertilizing manner, there exists opportunity for timely conjunctive reaction to changing interests and needs.[6] Exhibit 6.1 outlines the alliance operating structure and protocol.

The Africa-MNC strategic alliance relationship begins with adoption and sanctioning of the alliance arrangement by African countries – the African Union. Multinationals independently sanction the alliance by expression of interest in joining the alliance and ultimately conducting enterprise in African countries. NGO participation is acknowledged as the NGO indicates intent to work in partnership with the alliance. Periodic meetings are arranged to provide ongoing opportunity for policy discussions among alliance participants. Organizational control of the alliance passes to NEPAD as representative of the collective voice of African countries. The voices of the MNC and NGO are singularly represented initially but ultimately merge with NEPAD as NEPAD takes on oversight voice of the alliance. The merging of NEPAD and MNC input along with the NGO provides an apparatus for operational interface. The alliance becomes the NEPAD-MNC construct with NGO participation through which a formal working relationship is developed and cultivated. In framing the operational structure of the alliance, metrics are created to express ongoing impressions of alliance protocol. Satisfaction or dissatisfaction with alliance progress is expressed through this medium. Participants collaborate on appropriate resolutions including address and resolution of stakeholder issues; an inclusionary process provides opportunity for recognition and timely response. NEPAD's responsibility for alliance operations and oversight takes

---

[5] Sanyal, op. cit., p. 14.
[6] Ibid.

Exhibit 6.1 Africa-MNC Strategic Alliance Operating Structure and Protocol.

Alliance Adoption	African countries collectively sanction the alliance and hold periodic meetings Multinationals independently sanction the alliance and participate in alliance meetings NGO participation is acknowledged
Organizational Control	NEPAD represents collective voice of African countries MNC represents singular voice NGO represents voice of grassroots NEPAD represents oversight voice of the alliance
Operational/interface Apparatus*	NEPAD-MNC develop and cultivate suitable working relationship with NGO input Alliance creates metrics to express ongoing satisfaction/dissatisfaction with alliance protocol NEPAD-MNC identify differences and collaborate on resolution with NGO input NEPAD-MNC identify respective stakeholder issues and manage resolve with NGO input NEPAD responsible for alliance operations within agreed upon MNC and NGO input and participation.
Organizational Support	NEPAD develops ongoing financial support structure for alliance operations NEPAD and MNCs contribute financially to alliance pursuant to financial support structure
Market Mechanism	Alliance recognizes market forces as primary determinant of FDI, i.e., issues related to MNC type and location

*See Jonathan Hughes and Jeff Weiss, "Simple Rules for Making AlliancesWork," *Harvard Business Review*, November 2007.

place within agreed upon MNC and NGO input and participation. As the alliance organization requires financial support, a structure for this purpose is developed by NEPAD calling for NEPAD and MNC financial contribution (see FDI Fund in Chapter 7). As MNC participation expands, financial support expands as well thereby providing a sound financial structure for the alliance. The alliance recognizes that MNC participation is predicated upon the operation of market forces and therefore that FDI type and location are market determined.

6.1 Opportunistic Behavior

Ordinarily, as independent actors, African countries, multinational corporations and non-governmental organizations have the ability to behave opportunistically.[7] With no particular commitment to each other, each may operate solely in its own best interest potentially creating an atmosphere of mistrust. As partners in the alliance, mistrust or fear of partner insincerity or corrupt practices is minimized thus giving the alliance a more secure bond. On balance, one actor would not be able to take advantage of another. This is particularly important given the sometimes unflattering history of MNC-host country relations, particularly in the developing country arena. The sheer size of many MNCs and the valuable economic service they offer allows advantage-taking from time to time. The power of the multinational is particularly overwhelming in business arrangements as it is able to bargain for desired accommodations and negotiate the terms of the arrangements in its favor.[8]

On the other hand, host governments without oversight in certain instances display peculiar behavior in business dealings – believed to be acts of deception and corruption. Since business arrangements are often for extended periods of time, there exists the potential for renegotiation of original arrangements. Further, from time to time multinationals encounter charges of economic exploitation by domestic political opposition groups who generate enormous pressure to restructure business arrangements in favor of the host country.[9] As for NGOs, recall Sanyal's assessment that they tend to be extremely competitive, rarely form institutional linkages among themselves and that each favors the approach of convincing donors that it alone successfully undertakes grassroots projects.[10]

---

[7] Ilgaz Arikan and Oded Shenkar, "National Animosity and Cross-Border Alliances," *Academy of Management Journal*, 56:6, December 2013, pp. 1516–1544; Y. Luo, "Are Joint Venture Partners More Opportunistic in a More Volatile Environment? *Strategic Management Journal*, 28(1), 2007, pp. 39–60; J.J. Reuer, M. Zollo, and H. Singh, "Post-formation Dynamics in Strategic Alliances," *Strategic Management Journal*, 23(2), 2002, pp. 135–151; and J.E. Oxley, "Appropriability Hazards and. Showing Governance in Strategic Alliances: A Transaction Cost Approach," *Journal of Law, Economics & Organization*, 13(2), 1997, pp. 387–409.

[8] See a discussion of MNC-host business arrangements and bargaining models in Bobo, op. cit., pp. 118–121.

[9] Ibid.

[10] Sanyal, op. cit., p. 9.

With the collective authority to monitor African states, to interpret policy for the collective African nation-states partnership, and to negotiate arrangements supporting the partnership's objectives, oversight of opportunistic behavior by African countries falls within the purview of NEPAD. Hence the multinational is less exposed to the atomistic environment occasioned by doing business with potentially fifty-five African countries each potentially behaving opportunistically especially when corruption is the prickly adversary. At the same time, the power of the multinational to behave opportunistically or to overwhelm individual nation-states acting independently in negotiating business arrangements is largely minimized as NEPAD literally exhibits as much or more power than the MNC in that it represents the collective power of fifty-five African nations. Importantly, the power of NEPAD to behave opportunistically is minimized as well, as the MNC has the wherewithal to hold its own in dealings with NEPAD despite its formidable power base. After all, opportunity for development support including new capital formation, technology transfer, import substitution, hard currency infusion and the like remain very attractive incentives for NEPAD cooperation. And, in its augmenting role, the NGO is beholding to the alliance and must abide by its desires. That "the bottom-up approach was deliberately designed to discourage economic linkages between the top and the bottom..."[11] allowing NGOs greater autonomy to behave opportunistically, the alliance brings NGO activity under its auspices whereby the NGO neither has the resources nor the power to pursue opportunism especially where NEPAD and the MNC are concerned. Thus, in the Africa-MNC alliance, any power imbalance is largely minimized hence chance for opportunistic behavior minimized as well.

6.2 Opportunistic Balancing

Such inner working of the alliance effectively amounts to opportunistic balancing. Thus, in the context of the Africa-MNC strategic alliance, opportunistic balancing on the one hand mitigates the argument that the MNC engages the host simply to get what it wants. And on the other hand, the argument that the host country's domestic politics impact the bargaining process is mitigated as well. Further, linking these actors with the NGO under the alliance structure diminishes the notion that such linkage provides a conduit for "exploitative

---

[11] Ibid., p. 8.

122   *Africa-MNC Strategic Alliance*

surplus extraction from the bottom to the top."[12] This goes to validation of the alliance relationship. Moreover, opportunistic balancing speaks to the inner sanctum of capital theory, outlined in Chapter 2, in which competition drives efficiency. Opportunistic balancing more ensures transparent competition and fair play between multinationals and NEPAD with the strength of validating a family of interlinking paradigms giving structure to multinational enterprise as predicated by constructs such as the capitalist model, Western finance model, shareholder wealth maximization model, trade theory model, among others.[13]

These models have given rise to much debate regarding the efficacy of multinational enterprise in Africa. All characterizing the pursuit of private wealth and ownership of the means of production, multinationals are tasked with, voluntarily or involuntarily, demonstrating the utility of such paradigms in redressing the capacity innovation conundrum in the developing world. To be sure, multinational enterprise is no stranger to Africa; much has been documented about the relationship and much of that striking a sour note. Hence, much skepticism surrounds multinational enterprise on the African continent as many fear the power of the multinational corporation. But rather than rejecting the multinational's power, why not harness it. Rather than depriving Africa of the tremendous aid to capacity innovation that multinational enterprise may offer, why not seize the opportunity to enter into an alliance with the multinational. Overcoming the fear of multinational power may require little more than seeking the appropriate resolve; the Africa-MNC strategic alliance and opportunistic balancing offer such resolve.

6.3 Developing a Mindset

Recognizing that the strategic alliance and opportunistic balancing concepts in ways challenge the imagination and indeed face the possibility of being rejected out of hand, a process of developing a mindset for the concepts is proposed that suggests a method of preparing and acclimatizing actors and constituents to enable operationalizing and actualizing the concepts as outlined in Exhibit 6.2. This process takes place through three actions.[14] Taking measures to develop a clear 'tolerance' for multinational enterprise as a capacity innovation change agent sets the process in motion. This involves adopting a liberal attitude and open mind toward a direct and definitive

---

[12]Ibid.
[13]James Fulcher, *Capitalism*, 1st ed., New York: Oxford University Press, 2004.
[14]It is not the intent here to suggest that the process is necessarily linear.

6.3 Developing a Mindset   123

Exhibit 6.2 Developing a Mindset Operationalizing and Actualizing the Alliance and Balancing Concepts.

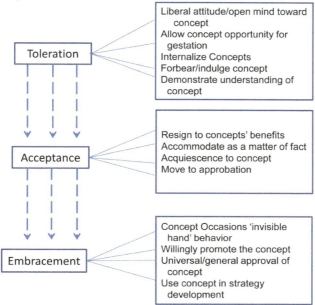

relationship between Africa and the multinational along with the NGO. Introduction of new concepts to the polity is often an onerous undertaking that may require sufficient time for consideration and examination. Conventional wisdom suggests allowing the concepts opportunity for gestation. Actors and constituents need time to internalize new ideas and to develop a sense of bonding. A measure of forbearance and indulgence by the actors and constituents facilitates the bonding process. As the process unfolds and examination of the concepts becomes more intense, the ability to demonstrate understanding of the concepts becomes more pronounced. As the initial action moves forward, toleration for the concepts takes hold therewith setting in motion the second action.

Following toleration, the process of 'acceptance' of the concepts begins to unfold. In so doing, actors and constituents resign themselves to the concepts' benefits. Much is potentially at stake: foreign direct investment and accompanying capacity innovation for the state; natural resource supply, new markets and shareholder wealth maximization opportunities for the multinational; and grassroots accommodation for the NGO. As a practical matter, all involved are in line for accommodation; recipients of the concepts' benefits improve their status by receiving and/or delivering economic gains. With accommodation as incentive, the acceptance process continues with acquiesce to the concepts and concludes by moving to approbation. In light of the potentially enormous benefits at stake, approval of the concepts – while not assured – is a rational decision.

And lastly the process moves to "embracement" of the alliance and opportunistic balancing concepts as viable tools for facilitating capacity innovation on the continent. In this instance, the expectation is that the alliance and opportunistic balancing mechanisms occasion forces that emulate the "invisible hand" as the actors and constituents willingly promote the concepts through reliance on market forces and market transactions to allocate resources and select "winners and losers." As change agent the multinational prompts this activity through FDI across the continent; the state encourages and cultivates FDI in continent-wide capacity innovation strategy development; and the NGO targets broad-based community innovation for grassroots constituents. There is essentially a mergence of the actions creating interplay of sorts whereby toleration, acceptance and embracement of the concepts result throughout the polity. A mindset for the Africa-MNC strategic alliance and opportunistic balancing concepts materializes and holds sway throughout the continent. The formal structure for the strategic alliance and opportunistic balancing process is articulated in forthcoming dialogue in Chapter 7.

6.4 The Corruption Dilemma

The Africa-MNC alliance amounts to Africa taking charge of its own affairs. Embracing the alliance signals Africans are undertaking a path of collective self-reliance orchestrated by their own internal mechanism. However, success in this venture may require Africa to resolve a deeply vexing issue that is seemingly characteristic of the African lifestyle. The issue, corruption – "the abuse of the public trust or the use of political influence for personal gain at the public's cost is widespread and deeply rooted in" Africa. Corruption is arguably the single most image tarnishing and capacity innovation delimiting factor in Africa's pursuit of a recognized global presence. "It has compromised the whole of the institutional mapping in some cases and seriously constricted it in others. The constitution is little more than words on paper, the policy apparatus is bypassed, institutional capacity is impaired, political feasibility is violated, and government capacity is restricted. These conditions lead to undermined property rights, rule of law, investment incentives, development objectives, and the human development paradigm. Corruption is quite literally an assault on the political, social, and economic well-being of a nation. It creates systemic inefficiencies that severely restrict development and modernization. Key public officials and principal members of civil society from more than 60 developing countries ranked public sector corruption as the most severe impediment to development and growth."[15]

There has long been the impression that corruption on the continent is beyond the reach of "good governance," "the rule of law" and certainly "ethics." But under the proposed relationship between Africa and the multinational corporation, corruption falls within the domain of the alliance and opportunistic balancing. Any unscrupulous behavior, particularly which pejoratively affects the alliance, falls under the oversight of NEPAD including the opportunistic balancing apparatus. Quite frankly, the stage is set by the Heads of State and Government of the Member States of the African Union in instructing the African Union to "promote and protect human and peoples' rights in accordance with the African Charter on Human and Peoples' Rights and other relevant human rights instruments"[16] and through this African Union preambles that provide a platform and structure through NEPAD doctrines and political governance initiatives that arguably extend the long arm of the African Union to corruption oversight and mitigation. This is apparent through charges prescribed and mandated for NEPAD. The following excerpts so attest:[17]

- Across the continent, Africans declare that we will no longer allow ourselves to be conditioned by circumstance. We will determine our own destiny and call on the rest of the world to complement our efforts. There are already signs of progress and hope. Democratic regimes that are committed to the protection of human rights, people-centered development and market-oriented economies are on the increase. African peoples have begun to demonstrate their refusal to accept poor economic and political leadership. These developments are, however, uneven and inadequate and need to be further expedited.

---

[15] Bobo, op. cit., pp. 211–212. See also pages 200–201 in this source for a discussion of the human development paradigm. Original narrative on the human development paradigm is presented in Mahbub ul Haq, *Reflections on Human Development*, Oxford: Oxford University Press, 1995, pp. 228–229.

[16] Constitutive Act of the African Union, Adopted by the Thirty-Sixth Ordinary Session of the Assembly of Heads of State and Government, Lome, Togo, July 11, 2000, http://www.au.int/en/sites/default/files/ConstitutiveAct_EN.pdf.

[17] The New Partnership for Africa's Development (NEPAD), October 2001, #7, #79–#85, #180, and #185.

- Promoting and protecting democracy and human rights in their respective countries and regions, by developing clear standards of accountability, transparency and participatory governance at the national and subnational levels ...
- It is generally acknowledged that development is impossible in the absence of true democracy, respect for human rights, peace and good governance. With the New Partnership for Africa's Development, Africa undertakes to respect the global standards of democracy, the core components of which include political pluralism, allowing for the existence of several political parties and workers' unions, and fair, open and democratic elections periodically organized to enable people to choose their leaders freely.
- Contribute to strengthening the political and administrative framework of participating countries, in line with the principles of democracy, transparency, accountability, integrity, respect for human rights and promotion of the rule of law ...
- The states involved in the New Partnership for Africa's Development will also undertake a series of commitments towards meeting basic standards of good governance and democratic behavior while, at the same time, giving support to each other. Participating states will be supported in undertaking such desired institutional reforms where required. Within six months of its institutionalization, the leadership of the New Partnership for Africa's Development will identify recommendations on appropriate diagnostic and assessment tools, in support of compliance with the shared goals of good governance, as well as identify institutional weaknesses and seek resources and expertise for addressing these weaknesses.
- In order to strengthen political governance and build capacity to meet these commitments, the leadership of the New Partnership for Africa's Development will undertake a process of targeted [capacity innovation] initiatives. These institutional reforms will focus on ... adopting effective measures to combat corruption and embezzlement ...
- ... Set up coordinated mechanisms [*with Africa by industrialized countries and multilateral organizations*] for combating corruption effectively, as well as committing themselves to the return of monies (proceeds) of such practices to Africa.

These decrees clearly signal the AU's predisposition to erecting democracy, good governance and the rule of law on the continent with a particular focus on tackling human rights violations and corruption. The AU's power to enforce the decrees rests in the *Protocol Relating to the Establishment of the Peace and Security Council of the African Union* which charges the African Standby Force, comprised of military, police and civilian capabilities, with the duty to: promote and encourage democratic practices, good governance and the rule of law, protect human rights and fundamental freedoms, respect for the sanctity of human life and international humanitarian law... and further... the African Standby Force shall ... perform functions ... as may be mandated by the Peace and Security Council or the Assembly.[18] The chain of authority – the African Heads of State to the African Union to The New Partnership for Africa's Development to the African Standby Force, articulates a general administrative structure important to enabling the conduct of the wishes of the African Heads of State.

6.5 Corruption, Human Rights Violations and Crimes Against Humanity: Intersecting Acts

The charges laid upon NEPAD with the African Standby Force as backup underscore not only the AU's authority to engage corruption and human rights violations but the political will and enforcement apparatus to effect action. As human rights violations, perhaps arguably, encompass the very worrisome problem of corruption, placing them under NEPAD's jurisdiction is a good policy decision. The AU acknowledges that capacity innovation is impossible in the absence of true ... respect for human rights. NEPAD's efforts to carry out capacity innovation – its main objective, take on an additional measure of control by being able to directly engage acts of corruption as human rights violations. Support for this approach comes from the Supreme Court of India in a 2012 judgment which held that "corruption ... undermines human rights, indirectly violating them," and that "systematic corruption is a human rights violation in itself."[19] Further, the Constitutional Court of South Africa asserted that "[c]orruption and maladministration are inconsistent with the rule of law and the fundamental values of our Constitution. They undermine the constitutional commitment to human dignity, the achievement of equality and the advancement of human rights and freedoms."[20] And, Ahmed argues that

---

[18] See Appendix: Protocol Relating to the Establishment of the Peace and Security Council of the African Union, Article 4 and Article 13.

[19] Quote found in Anne Peters, *Corruption and Human Rights*, Working Paper Series No. 20, Basel Institute on Governance, September 2015, p. 12.

[20] Quote found in Anne Peters, ibid.

"the struggle to promote human rights and the campaign against corruption share a great deal of common ground since they intersect . . . each other."[21] He asserts that "it is accepted and agreed that corruption is injurious to public administration, undermines democracy, degrades the moral fabrics of the society and violates human rights and social fabric of the society."[22] The International Council on Human Rights Policy, in the report *Corruption and Human Rights: Making the Connection*, offers an informative narrative on acts of corruption as human rights violations arguing that: "A clear understanding of the practical connections between acts of corruption and human rights may empower those who have legitimate claims to demand their rights in relation to corruption, and may assist states and other public authorities to respect, protect and fulfill their human rights responsibilities at every level."[23]

The report further informs that "Some international documents have even considered corruption to be a crime against humanity."[24] This is a particularly interesting notion as the AU's operating principles provide "the right of the Union to intervene in a Member State pursuant to a decision of the Assembly in respect of grave circumstances, namely: . . . crimes against humanity . . ."[25] It is this element that conceivably provides an opportunity for the AU to intervene in the affairs of a sovereign state regarding acts of corruption. While corruption and human rights violations as having common ground bears truth, does corruption rise to the level of crimes against humanity? The reply here logically begins with the meaning of crimes against humanity. Inhumane acts committed as part of a widespread or systematic attack directed against civilian populations is commonly cited as crimes against humanity.[26]

---

[21] Barkat Aftab Ahmed Khalid Ahmed, "Role of Corruption in Human Rights Violation," *Scholarly Research Journal for Humanity Science & English Language*, Vol. I, Issue IV, June–July 2014, p. 643.

[22] Ibid.

[23] *Corruption and Human Rights: Making the Connection*, International Council on Human Rights Policy, Versoix, Switzerland, 2009, p. 5.

[24] Ibid., p. 23.

[25] Constitutive Act of the African Union, op. cit.

[26] *Convention on the Non-Applicability of Statutory Limitations to War Crimes and Crimes Against Humanity*, Office of the High Commissioner for Human Rights, New York: United Nations, November 11, 1970, http://www.ohchr.org/EN/ProfessionalInterest/Pages/WarCrimes.aspx; Stephanie Hanson, *Corruption in Sub-Saharan Africa*, Council on Foreign Relations, Washington, DC, August 6, 2009; and Dennis Wanyenji Njoroge, *Grand Corruption as a Crime Against Humanity*, http://kenyalaw.org/kl/fileadmin/pdfdownloads/Moi_B_RESEARCH.pdf.

In this context, viewing corruption as multi-faceted, so to speak, allows greater ease in assessing the extent of "widespread or systematic" and "attack directed against" that scales them, in the context of corruption, to the level of crimes against humanity. Fundamentally, corruption has three strands: "petty corruption involving everyday abuse of entrusted power by low- and mid-level public officials in their interactions with ordinary citizens, who often are trying to access basic goods or services in places like hospitals, schools, police departments and other agencies; grand corruption consisting of acts committed at a high level of government that distort policies or the central functioning of the state, enabling leaders to benefit at the expense of the public good; and political corruption concerning manipulation of policies, institutions and rules of procedure in the allocation of resources and financing by political decision makers, who abuse their position to sustain their power, status and wealth."[27]

Invariably, governments are victims of petty, grand and political corruption; all acts that harm a society from low-level bribes to high-level political graft. Grand and political graft impose the largest financial cost on a society – on the scale of millions of dollars, but petty bribes to police officers, customs officials and the like "have a corrosive effect on basic institutions and undermine public trust in the government."[28] Transparency International asserts that "corruption impacts societies in a multitude of ways. In the worst cases, it costs lives. Short of this, it costs people their freedom, health, or money. The cost of corruption can be divided into four main categories: political, economic, social, and environmental. On the political front, corruption is a major obstacle to democracy and the rule of law. In a democratic system, offices and institutions lose their legitimacy when they're misused for private advantage. This is harmful in established democracies, but even more so in newly emerging ones. It is extremely challenging to develop accountable political leadership in a corrupt climate. Economically, corruption depletes national wealth. Corrupt politicians invest scarce public resources in projects that will line their pockets rather than benefit communities, and prioritize high-profile projects such as dams, power plants, pipelines and refineries over less spectacular but more urgent infrastructure projects such as schools, hospitals and roads. Corruption also hinders the development of fair market structures and distorts competition, which in turn deters investment. Corruption corrodes

---

[27]Transparency International (the global coalition against corruption), http://www.transparency.org/whoweare/organisation/faqs_on_corruption.

[28]Hanson, op. cit., p. 2 and Njoroge, op. cit.

the social fabric of society. It undermines people's trust in the political system, in its institutions and its leadership. A distrustful or apathetic public can then become yet another hurdle to challenging corruption. Environmental degradation is another consequence of corrupt systems. The lack of, or non-enforcement of, environmental regulations and legislation means that precious natural resources are carelessly exploited, and entire ecological systems are ravaged. From mining, to logging, to carbon offsets, companies across the globe continue to pay bribes in return for unrestricted destruction."[29] And as UNODC Executive Director Antonio Costa admonished: "Corruption kills: literally, through counterfeit medicines; by bribes paid to security officials that enable terrorist attacks. Corruption kills trust – in government, public institutions, and companies. It kills the environment – through the dumping of hazardous waste, illegal logging, over-fishing, or the extraction of blood diamonds. And corruption kills growth – by stealing public money needed for schools, hospitals, roads, and by driving business into the shadow economy."[30]

The weight of corruption may be seen in its pejorative impact on a country's economy and through this a country's people. Clearly corruption is a bane on society perpetuated by public officials set on personal gain at the public's expense. The pugnaciousness of such behavior raises the specter of inhumane acts that give rise to crimes against humanity. It is important to point out that only the "widespread" or the "systematic" marker along with the "attack directed against" marker, mentioned above, need to be satisfied to raise corruption to the level of crimes against humanity.[31] Assessment of a country's corruption score provides a very salient starting point from which to gauge these markers. The corruption score is provided by Transparency International's Corruption Perceptions Index – a global indicator of public sector corruption, offering a yearly snapshot of the relative degree of corruption by ranking countries from all over the globe.[32] Exhibit 6.3 shows corruption scores for African countries and indicates the extent of corruption on a scale of 0 (highly corrupt) to 100 (very clean). Quite apparently, the continent is literally mired in corruption. No African country scores higher than

---

[29]Transparency International, op. cit.

[30]Antonio Maria Costa, *Anti-Corruption Climate Change: It Started in Nigeria*, United Nations Office on Drugs and Crime, 6th National Seminar on Economic Crime, Abuja, Nigeria, November 13, 2007, http://www.unodc.org/unodc/en/frontpage/nigerias-corruption-busters.html

[31]Njoroge, op. cit.

[32]Transparency International, op. cit.

6.5 Corruption, Human Rights Violations and Crimes Against Humanity

Exhibit 6.3 African Countries 2015* Corruption Score.

Botswana	63	Gabon	34	Uganda	25
Rwanda	54	Niger	34	Central African Republic	24
Mauritius	53	Ethiopia	33	Congo (Republic)	23
Namibia	53	Cote d'Ivoire	32	Democratic Republic of the Congo	22
Ghana	47	Togo	32	Burundi	21
Senegal	44	Mauritania	31	Zimbabwe	21
South Africa	44	Mozambique	31	Eritrea	18
Sao Tome & Principe	42	Tanzania	30	Guinea-Bissau	17
Burkina Faso	38	Sierra Leone	29	Libya	16
Tunisia	38	Gambia	28	Angola	15
Zambia	38	Madagascar	28	South Sudan	15
Benin	37	Cameroon	27	Somalia	8
Liberia	37	Comoros	26		
Algeria	36	Nigeria	26		
Egypt	36	Guinea	25		
Morocco	36	Kenya	25		
Mali	35				
Djibauiti	34				

*A country's score indicates the perceived level of public sector corruption on a scale of 0 (highly corrupt) to 100 (very clean).

Source: Corruption Perceptions Index 2015, Transparency International, www.transparency.org/cpi2015

63 and most fall below 38; the average score is 31. Further, "Academic research shows that a one-point improvement in a country's Transparency International corruption score is correlated with a productivity increase equal to 4 percent of gross domestic product (GDP)."[33] With six of the world's ten most corrupt countries located in Africa, a 2002 African Union report "estimated that corruption costs African economies in excess of US$148 billion a year. Both direct and indirect costs of corruption not only represent 25 percent of Africa's GDP but also increase the cost of goods by as much as 20 percent. This is how corruption has retarded development of the continent by undermining indigenous entrepreneurship and scaring away foreign investments."[34] In a study of corruption in Nigeria, PricewaterhouseCoopers (PwC) estimates that the nation alone lost roughly $185 billion in GDP over the last fifteen

---

[33] Hanson, op. cit., p. 2.
[34] Wafula Okumu, *The Role of AU/NEPAD in Preventing and Combating Corruption in Africa – A Critical Analysis,* http://www.africafiles.org/printableversion.asp?id=10150. See also Hanson, op. cit., p. 1.

years as corruption lowered tax revenue, discouraged investment – especially FDI, and reduced human capital – particularly associated with technology transfer. The international auditing and consulting firm projects corruption could cost Nigeria more than a third of GDP by 2030 if left to its own devices. This cost equates to around $1000 per Nigerian today and nearly $2000 by 2030.[35] Numerous cases of large-scale corruption by public officials have been reported in Zimbabwe in recent years; "the fraud department of the Zimbabwe Republic Police reported that 91 percent of the cases it investigated in 1998 had occurred in the government, and three-quarters of them had involved the award of tenders and contracts. In 1999–2001, misuse of resources in the Zimbabwean government and in state-run companies cost the country close to $800 million."[36] Further, the African Development Bank in collaboration with Global Financial Integrity reported that "illicit financial flows from Africa amounted to $1.4 trillion between 1980 and 2009 – more than the economic aid and foreign direct investment the continent received during that period."[37] And in a rather capstoning comment regarding corruption in Africa, former U.S. President Barack Obama joined African leaders in expressing the need to "step up our collective efforts against the corruption that costs African economies tens of billions of dollars every year – money that ought to be invested in the people of Africa."[38] These findings frame the narrative for the whole of Africa: corruption is ubiquitous – no country is spared, methodical – its effect upon the citizenry is inescapable and, inclusive – the whole of civil society is affected.

It is clear at this point that the cost of corruption to the continent is on a grand scale; certainly of a magnitude sufficient to raise the matter to the level of crimes against humanity. But before concluding the issue, Njoroge offers further perspective on acts of corruption as crimes against humanity. Deliberating on the matters of "widespread" and "systematic" he argues that "The concept of 'widespread' as an element of crimes against humanity

---

[35] See Shannon K. O'Neil, *This Week in Markets and Democracy: Moldova's Protests, Investors take on Graft, Corruption's Cost in Nigeria*, Council on Foreign Relations, February 5, 2016; and *Impact of Corruption on Nigeria's Economy*, PricewaterhouseCoopers, (PwC), *2016*, http://www.pwc.com/ng/en/publications/impact-of-corruption-on-nigerias-economy.html

[36] Nsongurua J. Udombana, "Fighting Corruption Seriously? Africa's Anti-corruption Convention," *Singapore Journal of International & Comparative Law*, 2003, Vol. 7, p. 450.

[37] "Extracting Justice: Battling Corruption in Resource-Rich Africa," *Devex*, August 14, 2014, https://www.devex.com/news/extracting-justice-battling-corruption-in-resource-rich-africa-84137

[38] Ibid.

6.5 Corruption, Human Rights Violations and Crimes Against Humanity

was introduced in the 1990s as an accepted formulation for the contextual threshold, thus contributing to clarity and consistency in this area of law and therefore an attack requires at least some minimal aspect of the two, widespread generally connotes the large scale of the attack and the number of the victims but it should be noted that no numerical limit has been set; the issue must be decided on the facts. Widespread is typically known to refer to the cumulative effects of numerous inhumane acts; it could also be satisfied by a singular massive act of extraordinary magnitude."[39] Njoroge adds that "... grand corruption attacks a wide number of victims consequently due to looting of national funds and thus deaths resulting from hunger and illnesses, this fact was acknowledged by the East Timor Truth Commission, which calculated that 84,200 of the 102,800 victims in the country died of hunger and illness rather than being killed outright or forcefully disappeared during the Indonesian occupation. Further The Corruption Perceptions Index [2015] serves as a reminder that the abuse of power, secret dealings and bribery continue to ravage societies around the world, South Sudan being highly corrupt with a score [12] thus further proving the widespread element when one considers the affected population in terms of what they will lack consequently due to corruption."[40]

On the matter of "systematic," Njoroge lends that the term has been defined variously as: organized nature of the acts or pattern of conduct; following a regular pattern; on the basis of common policy; involving substantial public or private resources; large scale or continuous commission; and implication of high level authority.[41] These elements suggest a broad range of collective acts of corruption involving systemic and sustained dishonest behavior by public officials. The final marker – "attack directed against," in connection with crimes against humanity refers to multiple prohibitive acts committed upon multiple victims or directed against the civilian population. Nothing in customary international law, Njoroge confides, requires proof of the existence of a plan or policy to commit such acts.[42] Commission of the acts on prima facie grounds is sufficient testament to the acts themselves.

So what conclusion may be reached regarding acts of corruption as crimes against humanity? While petty corruption arguably rises to the level of crimes against humanity, grand corruption and political corruption clearly reach the

---

[39] Njoroge, op. cit.
[40] Ibid.
[41] Ibid.
[42] Ibid.

mark. Instructively, the ubiquitous character of corruption signals that it is widespread; every African country feels its weight and painfully so. The methodical nature of corruption confers that it is systematic; no African can avoid its punitive demands. And the inclusiveness of corruption renders it an attack directed against all Africans; civil society in its entirety suffers the indignity of African public officials violating the trust of the people. In total, corruption – in its various forms, deprives the people of Africa – the children no less, of their right to be nurtured by the benefits of the continent's resources. Many entrusted with safeguarding the natural riches and returns from their development have instead enriched themselves for decades. Billions of dollars have been pilfered from Africa where they could have done much good and placed elsewhere where they are doing much good – but not for Africans. Nigeria serves as a poignant example: "Unscrupulous leaders pilfered the national coffers and stashed away billions of dollars in foreign bank accounts. By some estimates close to US$400 billion was stolen between 1960 and 1999."[43] The enormity of the problem is beyond reason – crimes by Africans against Africans; they are surely crimes against humanity.

6.6 Interdicting Corruption: The Trump Card

The alliance arrangement coupled with opportunistic balancing may effectively operate as Africa's "trump card" against corruption. With NEPAD charged as facilitator and coordinator of the development and implementation of continent-wide capacity innovation schemes, its actions hold a measure of implied power over decision-making authority traditionally held within the sovereign scope of nation states. This prevails despite the African Union's objective to defend the sovereignty, territorial integrity and independence of its member countries. However, if corruption is to be overcome in full or minimized to the point of ineffectiveness, NEPAD's power must be more than implied. The direct power to interdict corruption at its very origin is requisite and the African Union has the ability to so empower NEPAD. Recall that under the AU's operating principles, there exists: "the right of the Union to intervene in a Member State pursuant to a decision of the Assembly in respect of grave circumstances, namely: ... crimes against humanity . . ."[44] Further, "the Assembly may delegate any of its powers and functions to any organ of the

---

[43]*Nigeria's Corruption Busters*, United Nations Office on Drugs and Crime, November 20, 2007; and Antonio Maria Costa, op. cit.

[44]Constitutive Act of the African Union, op. cit.

Union."[45] Under these provisions, the matter to interdict corruption reduces to political authority versus political will. Having established that corruption is in fact a crime against humanity, the way forward is manifest. The African Union has the political authority to interdict; does the Union have the political will to do so?

A fully forward-looking corruption policy invariably requires placement of interdiction capability directly at the source of interdiction. Substantively this involves empowering NEPAD with the mandate to undertake means as necessary to rid the continent of corruption, including sanctioning power and even referring offenders to the Court of Justice of the Union; thereby putting in play the "trump card."[46] Any attempt to circumvent NEPAD's authority particularly through engaging in any form of corrupt practices by either an indifferent nation state – or parties thereto, or a multinational corporation would be subject to the incontestable override of the "trump card." Rogue nation states or multinationals not subscribing to policies of the alliance would be subject to the sanctioning power of NEPAD. This would work in one instance pretty much in a manner similar to actions taken by the "world order" of nations in reigning in countries seen as breaching general world order protocol. Such countries face certain and often harsh sanctions – usually of an economic nature but may include humanitarian measures as well. In other instances, actions take the form of those imposed upon members of, say, the WTO, regional trade pacts, military alliances, etc., for rules violations: trade partners face fines, revocation of preferential tariffs, etc.; military allies sustain interdiction of sovereign powers, among other proscriptions. In the Africa-MNC alliance, breaches of alliance protocol would prompt similar actions: member countries would suffer direct sanctions; multinationals would suffer indirectly as doing business with violators would be complicated by sanctions. For non-member countries, sanctions would apply as well and, in the worse-case scenario, non-members would become veritable outcasts. Such would be the hazards of Africa's trump card.

There is a clear message underlying the strategic alliance: Africa's role is at a minimum equal to that of the partners. Any notion that Africa is in search of a "handout" rather than a "hand shake" would find its absurdity in the Africa-MNC alliance. Africa's value to the alliance is indisputable. Its enormous wealth in natural resources, a colossal consumer base masking tremendous

---

[45] Ibid.
[46] This reference is not intended in any way or by any means to reflect the policies of U.S. President Donald J. Trump.

pent-up demand along with potentially cheap labor to support labor intensive FDI are more than warranted to collateralize its stake in the alliance. The Africa-MNC alliance would contravene historical Anglophone, Francophone, and the perhaps emerging Chinaphone branding. The Africa-MNC alliance signals a new African order – African branding. The new order would not be bounded by the African continent but would find its accommodation in the globalization process. The alliance and opportunistic balancing would so decree.

7
Path to Capacity Innovation: An Africa-MNC Strategic Alliance

The standalone approach to African capacity innovation discussed in Chapter 3, while offering an encouraging baseline assault on the matter, is not recommended. Africa simply lacks the independent wherewithal, the internal capacity or country capacity ID to mount a development structure – in the near future – sufficient to achieve take-off; the continent's plethora of shortcomings so attest. Without take-off, there can be no sustained innovation. Importantly, Africa pursues an 'aid for trade' scheme as a funding modality to facilitate capacity innovation, calling for international assistance to address capacity deficiencies and supply-side constraints to facilitate participation of African countries in international markets; to cope with transitional adjustment costs from trade liberalization; and to assist trade development, trade-related infrastructure, among other needs.[1] Three narratives by the African Union outline plans for transitioning the continent to sustained capacity innovation; the initial scheme outlining objectives, priorities and strategies, and two action plans detailing execution procedures.[2] None of the plans mentions "created" comparative advantage or multinational corporations as means of redressing capacity deficiencies and supply-side constraints concerns, certainly not to any significant extent. While the African Union's intent is to confront the full range of circumstances impeding Africa's capacity innovation efforts, there

---

[1] *Revision of the AU/NEPAD AFRICAN ACTION PLAN 2010–2015: Advancing Regional and Continental Integration Together through Shared Values*, Abridged Report 2010–2012, NEPAD Planning and Coordinating Agency, 2011, p. 41.

[2] See *The New Partnership for Africa's Development (NEPAD)*, October 2001; *The AU/NEPAD African Action Plan 2010–2015: Advancing Regional and Continental Integration in Africa*, African Union, October 17, 2009, and *Revision of the AU/NEPAD AFRICAN ACTION PLAN 2010–2015: Advancing Regional and Continental Integration Together through Shared Values*, Abridged Report 2010–2012, NEPAD Planning and Coordinating Agency, 2011.

is a missing link particularly in its approach to resolving capacity and market access constraints.

There is need for a strategic capacity innovation policy framework, a scheme or model no less, that extends the reach of 'aid for trade,' expands the boundary of funding modality, and along the way stimulates movement up the ladder of comparative advantage, redresses core challenges, and postulates a definitive endgame of economic modernity and ultimately resolution of the African capacity innovation quandary. Exhibit 7.1 confers such a framework, proposing a strategic alliance between African countries and the multinational

Exhibit 7.1 Africa – MNC Strategic Alliance A Policy Framework for African Capacity Innovation.

A Forward-Looking Business Model	
African Countries (NEPAD)	Multinational Corporations (MNCs)
• Institutional component: provides institutional input to attract FDI and support the market mechanism purposely to encourage top down capacity innovation and trade development. Operates within guidelines of constitutional prescripts, rule of law and bodies of government.	Market component: provides market mechanism input through FDI that stimulates top down capacity innovation and trade development supported by institutional input.
• World Bank/ IMF provide support structure for NEPAD objectives, e.g., infrastructure development and institutions building.	World Bank/ IMF provide support capacities that provide a nurturing environment for attracting FDI/MNC.
• NGOs play grassroots technical assistance role encouraged by NEPAD and supported by stakeholder givebacks and donor contributions.	NGOs play bottom up capacity innovation role supported by stakeholder givebacks.
• Facilitate capital diffusion through encouragement of MNC participation throughout the African continent, and development and implementation of uniform policies and practices and provision of oversight and monitoring of FDI and capital diffusion activities.	Pursue foreign direct investment and conduct capital diffusion throughout the African continent in accordance with needs of African countries and FDI policies outlined by NEPAD.
• Support "created" comparative advantage through development of uniform policies and practices to accommodate collective foreign direct investment projects.	Facilitate "created" comparative advantage through initiation and implementation of foreign direct investment projects.
• Promote collective capacity innovation founded upon collective self-reliance and creation of regional markets and trade flows to facilitate regional integration directed toward redressing the critical mass problem.	Support regional integration of markets through promotion of trade flows between and among African countries. Change agent to remedy the critical mass problem.

Path to Capacity Innovation: An Africa-MNC Strategic Alliance

Exhibit 7.1 Continued

• Promote economic complementarity through development of uniform policies and practices to encourage production of goods and services to supply the mutual needs of African countries. Establish cooperation linkages and best practices.	Facilitate development of economic complementarity through production of goods and services to supply the mutual needs of African countries through complex integration strategies and supply chain management.
• Promote development of price-maker status through development and implementation of collective policies and practices that encourage and support accommodating activities.	Facilitate development of price-maker status through capital diffusion and development of comparative advantage throughout the continent.
• Promote African telecommunications superhighway through development of policies and practices that encourage such FDI.	Support FDI in intra-continental communication technology projects to link African countries through an African telecommunications superhighway.
• Develop infrastructure plan and encourage support of the development of infrastructure to accommodate creation of comparative advantage, regional economic integration, economic complementarity and intra-continental communication.	Facilitate development of infrastructure to support creation of comparative advantage, regional economic integration, economic complementarity and intra-continental communication.
• Coordinate and synchronize convergence of NEPAD, MNC and NGO activities.	Facilitate convergence of NEPAD, MNC and NGO activities.
• Facilitate abatement of critical mass problem through encouragement of regional integration and development of policies and practices that encourage FDI and capital diffusion activities.	Abate critical mass problem through support of regional integration and pursuit of FDI and capital diffusion activities.
• Address transfer pricing/trade misinvoicing problem.	Facilitate correction of transfer pricing/trade misinvoicing problem.
• Develop policies and practices to encourage support of a foreign direct investment fund.	Support development and maintenance of a foreign direct investment fund to support new capital formation activities and support NEPAD Operations.
• Develop NEPAD commercial bank to provide working capital and short-term loans to enterprises particularly serving the needs of multinationals.	Support NEPAD bank through MNC deposits and operating fees.
• Create linkage and operating protocol with African security force for FDI protection.	Interface with African security force and operate under African security force protocol for FDI protection.

(Continued)

Exhibit 7.1 Continued	
A Forward-Looking Business Model	
African Countries (NEPAD)	Multinational Corporations
• Facilitate employment of country capacity ID attributes.	Support employment of country capacity ID attributes.
• Facilitate organizational tie-in	Support organizational tie-in
• Facilitate unlocking the lock; promote African participation in globalization and global resource allocation. Endgame: Facilitate realistic near-future developed-country status for African continent.	Support unlocking the lock; enable expansion of African input in globalization and global resource allocation. Endgame: Support/enable near-future developed-country status.

corporation with input from the NGO and couched upon the NEPAD-MNC-NGO cross-fertilizing integrative structure articulated in Chapter 4.[3] The Alliance is the foundation upon which capacity innovation is constructed. The actors in the alliance working in partnership and in concert provide principal impetus for efficient interaction of salient features of the capacity innovation scheme. Traversing the ladder of comparative advantage and tackling the core challenges require a definitive roadmap, so to speak, that identifies a point of initiation, inputs and outcomes, and a clearly defined endgame. The NEPAD-MNC-NGO capacity innovation arrangement articulates such an undertaking; the forthcoming dialogue discusses important matters essential to the arrangement and prescribes the process of actualizing it. All sown, capacity innovation is the reward.

Creation of the capacity innovation policy framework, as depicted in Exhibit 7.1, begins with the recognition that the Africa – MNC Strategic Alliance anticipates two essential drivers: an institutional mapping characterized by good governance and efficient government institutions, and a free market philosophy characterized by the "invisible hand" and supported by political will. Unlike the standalone construct deliberated in Chapter 3, the Alliance foresees utilization of not just the collective power of African nations but also the full capacity of one of the most powerful entities in capacity innovation – the multinational corporation. Successful linking of

---

[3] Bobo presented an earlier version of this framework in *Neo-Liberalism, Interventionism and the Developmental State: Implementing the New Partnership for Africa's Development*. Exhibit 7.1 is an evolution of that dialogue – substantially so – utilized here to assist expression of the capacity innovation process as judged appropriate for Africa's predicament. See Benjamin F. Bobo and Hermann Sintim-Aboagye, eds., *Neo-Liberalism, Interventionism and the Developmental State: Implementing the New Partnership for Africa's Development*, Trenton, New Jersey: Africa World Press; 2012, pp. 228–231.

Africa and the MNC in an alliance arrangement with NGO participation creates a powerful structure for capacity innovation.

7.1 The Linking Process

In the pursuit of capacity innovation, importantly sustained capacity innovation, the ultimate interest is to speed movement towards modernity – as apparent in the First World – which concurrently prompts movement up the ladder of comparative advantage during which time an advanced market economy materializes, all directed toward the endgame of economic modernity and sustained capacity innovation. The linking of a series of components and activities occasioned by the 'invisible hand' (market prevails) and supported by political will (institutional mapping) gives rise to endgame success. In the Alliance, NEPAD plays the role of political will and the multinational casts as the market. The manner of appearance of components and activities in the framework does not necessarily suggest any specific order or sequence. The interaction of the various factors may take a number of forms but, for example, may play out in the following manner.

7.1.1 Institutional Component

On the country side, with the backing of African nations NEPAD interfaces with multinationals to stimulate top down capacity innovation – "the twist" – and trade development through promotion, facilitation and attainment of capacity innovation by way of encouraging foreign direct investment and capital diffusion accompanied by transmission, absorption and assimilation of technology. In so doing, NEPAD promotes an African institutional mapping that supports FDI, assures limited interference with the market mechanism, affirms observance of the democratic process, and pledges full support of government institutions to encourage an efficiently operating alliance. These efforts are in keeping with NEPAD principles: priority should be given to capacity innovation; private foreign finance is essential; market-oriented economies, democratic regimes, good governance and the rule of law are requisite; and institution building to support NEPAD principles is imperative. And further, the efforts support the AU/NEPAD African Action Plan to "improve the institutional framework for regional environmental governance."[4] It is useful

---

[4] *Revision of the AU/NEPAD AFRICAN ACTION PLAN 2010–2015: Advancing Regional and Continental Integration Together through Shared Values*, Abridged Report 2010–2012, NEPAD Planning and Coordinating Agency, 2011, p. 44.

to note here that Chang explored the development of different institutions of good governance in the history of developed countries and found that a number of factors currently considered preconditions for development were instead consequences of it – noting that much economic development occurred long before countries had thoroughly institutionalized democracies, corporate governance rules, and the like. He reasons that "Given that institutions are costly to establish and run, demanding [that developing countries] adopt institutions that are not strictly necessary can have serious opportunity cost implications.... Even when we agree that certain institutions are 'necessary,' we have to be careful in specifying their exact shapes."[5]

NEPAD in the state role here may raise concerns in light of two competing paradigms that offer perspectives on state policies and foreign investment decisions. On the one hand, neoclassical trade theory prescribes a reduced role for the state as it touts the self-sufficiency of the market. On the other hand, dependency theory – being decidedly more distrustful of multinational undertakings, holds the view that state institutions become hostage to foreign capital. But the experience of Singapore runs counter to both paradigms. Nizamuddin informs that the participation of multinational corporations in Singapore development has followed an upward trend despite government's adoption of numerous measures that reduced market risks and provided special incentives to foreign investment. The state assumed the lead in stimulating growth while at the same time managed input by powerful multinationals.[6] In this light, NEPAD's role is given special legitimacy particularly in the context of the Alliance where state leadership makes key input and the power of the multinational is moderated.

7.1.2 Market Component

On the MNC side, multinational corporations provide advanced form and intensity to the market mechanism, particularly from the top down "twist"

---

[5]Quoted in Merilee S. Grindle, "Good Enough Governance: Poverty Reduction and Reform in Developing Countries," Kennedy School of Government, Harvard University, (prepared for the Poverty Reduction Group of the World Bank), November 2002, p. 15; Ha-Joon Chang, "Institutional Development in Developing Countries in a Historical Perspective: Lessons from Developed Countries in Earlier Times." Unpublished paper, University of Cambridge, 2001, no page number.

[6]See Ali M. Nizamuddin, "Multinational Corporations and Economic Development: The Lessons of Singapore," *International Social Science Review*, 2007, Vol. 82, Issue 3/4, p. 149; Deepa Ollapally, "The South Looks North: The Third World in the New World Order," *Current History*, 92(573), 1993, pp. 175–179; and Don Walleri, "The Political Economy Literature on North-South Relations," *International Studies Quarterly*, 22(4), 589, 1978.

perspective. Enabled by their deep financial capacity and broad intellectual prowess, multinationals' undertaking of capital diffusion give rise to robust market operations. Quinlivan notes that "The economic role of multinational corporations (MNCs) is ... to channel physical and financial capital to countries with capital shortages. As a consequence, wealth is created, which yields new jobs directly and through "crowding-in" effects. In addition, new tax revenues arise from MNC generated income, allowing developing countries to improve their infrastructures and to strengthen their human capital. By improving the efficiency of capital flows, MNCs reduce world poverty levels and provide a positive externality . . ."[7]

Exhibit 7.2 outlines characteristics of the multinational, the immensity of which tells the story. Typically comprised of a number of subsidiaries with forward-looking objectives, the multinational corporation holds vast financial wealth and has ready access to capital markets and other financial arrangements. It can literally raise necessary capital to support a wide array of capital budgeting opportunities – such as FDI – and can negotiate advantages from lower taxes to accommodating workplace standards that augment

Exhibit 7.2 Characteristics of the Multinational Corporation.

- A large corporation comprised of many subsidiaries characterized by progressive posture.
- Financially wealthy with the ability to raise large sums of capital through the stock market and other financing opportunities.
- Strong negotiating powers in achieving locational advantages and other benefits that lower the cost of production, lower taxes, lower pollution standards, lower export-import tariffs, lower wages, flexible work hours, lower workplace standards.
- MNCs strive to maximize the value of stockholders' wealth and corporate profit.
- Primary strength in raw material conversion.
- Realizes considerable revenues from the production of consumer goods.
- Mobile: can relocate either in short run or long run to take advantage of new opportunities.
- Education rich: characterized by highly skilled workforce readily adaptable to technological changes and very capable of producing technology advances.
- Strong ability to initiate investment projects from the preliminary study to the economic and technological feasibility study to the engineering study.
- Strong management abililty.
- Access to and control of markets for movement of goods and services.
- Strong marketing skills.

Source: See Benjamin F. Bobo, *Rich Country, Poor Country: The Multinational as Change Agent*, Westport, CT: Praeger Publishers, 2005, pp. 202–203.

---

[7]Gary Quinlivan, "Sustainable Development: The Role of Multinational Corporations," Saint Vincent College, no date, p. 1.

cash flows and promote NPV > 0. Striving to maximize the wealth of its stockholders, the multinational takes full avail of location and internalization advantages[8] thus giving it primary strength throughout the entire product delivery process – from raw material extraction and conversion to production and marketing of consumer goods. Unlike the host country in which it locates, the multinational enjoys the benefit of being mobile. It can locate or relocate in a timely fashion to take advantage of new opportunities – a capacity that promotes stockholder wealth maximization and reduces risk particularly associated with market disturbances, nationalistic fervor, and the like. Its workforce is characteristically highly skilled and technologically savvy with capacities covering the entire spectrum of the ladder of comparative advantage. Importantly, underlying and predicate to the ladder of comparative advantage is a cadre of core workers who do the heavy lifting. Supported by innovations and technological advances, these workers virtually enable the capacity innovation process. Coupled with strong management and marketing competencies, MNC personnel engage FDI opportunities so astutely that access to and control of product markets are virtually assured. While these attributes bode well for the multinational, at the same time they provide rationale for pursuing FDI opportunities – benefits of which accrue to the host as well as the corporation. In an efficiently functioning alliance, MNC capacities transfer to the host's labor force and to the host's economic engine – through technology transmission, thereby inducing the capacity innovation process.

In the context of concerns about NEPAD, the multinational in the market role may raise concerns as well; foreign capital purportedly overpowers state institutions – leading to control of national economies. Peter Berger takes the view that foreign capital penetration of Third World economies is an exercise in growth without development. As he asserts: "'Growth without development' is based on the penetration of a Third World economy by foreign capital. This penetration results in a distortion of the economy, in the sense that it develops not in terms of international economic and social forces, but in the interest of the foreign 'metropolis.' ... The same 'distortion' creates an essentially colonial structure for the benefit of the foreign capitalists. It is not quite the same structure as that of the old colonialism, which was largely extractive, taking out raw materials from the colony that were needed for the industries in the home country. The new colonialism promotes industrialization, but

---

[8]Multinational enterprise derives from location specific advantages and firm specific advantages (internalization). See Benjamin F. Bobo, *Rich Country, Poor Country: The Multinational as Change Agent*, Westport, Connecticut: Praeger Publishers, 2005, pp. 175–176.

of a very peculiar kind. Generally it is capital-intensive rather than labor-intensive, thus actually creating unemployment in the 'developing' country.... Allocation of scarce resources to this kind of industrialization actually prevents development in other sectors of the society.... 'Neocolonialism,' therefore, implies increasing impoverishment and ever-greater dependency on foreign forces. In Andre Gunder Frank's graphic phrase, it means 'the development of underdevelopment'!"[9] In fact, Frank contends that developed countries – through multinational activity, have usurped the economic potential of Third World countries to support their own economic objectives.[10] But here too the Singapore case raises notable contradiction. State intervention by Singapore to reduce market risks has fostered development and special incentives to foreign investment have promoted sectoral development across the economy resulting in a general reduction in the level of poverty.[11] And, Quinlivan asserts that "Through free market initiatives, MNCs create wealth, which provides the income flow necessary for welfare improvements. If the desideratum of developing countries is to escape severe conditions of poverty, they need to privatize, deregulate, protect private property rights, and establish a rule of law – the MNCs will then provide the capital."[12]

In the Africa case, the structure of the Africa-MNC Strategic Alliance channels multinational prowess into productive output consistent with state objectives thereby containing the multinational's rather omnipresent power while at the same time assuring state sovereignty. The Alliance pursues a change to the status quo; it creates capacity whereby African priorities are the primary units of input, African market forces are the primary orchestrators of production and African people are the primary beneficiaries of output. Further, the Alliance is buttressed by the "code of conduct," – though voluntary, adopted by the Organization for Economic Cooperation and Development which includes standards for information disclosure, employee relations, taxation and national treatment.[13] In such a structure as tendered by the

---

[9]See Peter L. Berger, *Pyramids of Sacrafice*, New York: Basic Books, 1974, pp. 48–49.

[10]Ander Gunder Frank, "The Development of Underdevelopment," in David N. Balaam and Michael Veseth, Eds., *Readings in International Political Economy*, Upper Saddle River, NJ: Prentice-Hall, Inc., 1996, pp. 64–73.

[11]See Ali M. Nizamuddin, "Multinational Corporations and Economic Development: The Lessons of Singapore," ibid.

[12]Gary Quinlivan, "Sustainable Development: The Role of Multinational Corporations," Saint Vincent College, no date, p. 6.

[13]See Hans W. Singer and Javed A. Ansari, *Rich and Poor countries: Consequences of International Disorder*, 4th ed., Boston: Unwin and Hyman, 1988; and "Poor vs. Rich: A new Global Conflict," 1975, *Time*, December 22, pp. 34–35.

Alliance, foreign capital diffusion expressly reflects the role of the state in capacity innovation and the efficacy of the market in resource allocation.[14]

7.1.3 World Bank/IMF

The World Bank and International Monetary Fund (IMF) through their mandates provide vital support to African countries. This is particularly important to capacity innovation as such support, among other things, helps countries experiencing problems associated with capacity innovation. The World Bank provides loans, credits and grants to support a wide array of economic activity including investment in infrastructure and private sector development projects. The IMF provides support in the form of loans and technical assistance designed to help countries restore macroeconomic stability and develop their economies by upgrading institutions, rebuilding international reserves, stabilizing currencies, and paying for imports.[15] All are necessary conditions for capacity innovation and certainly facilitate attraction of foreign direct investment; upgraded African institutions and infrastructure particularly meet the call of the Alliance.

NEPAD's judicious pursuit of World Bank and IMF support in promoting its objectives and principles would serve well the Alliance's interests. Well-functioning institutions and advanced infrastructure provide fertile ground for FDI. However, it should be noted that African countries have had a checkered history with these institutions. Critique has revealed rather polarized views about the effect of World Bank and IMF economic programs on capacity innovation in targeted African countries. "Opinions range from criticisms of application of 'one-size-fits-all' models with almost predictable traumatic effect on the cost of living of the targeted country's populace, an over-emphasis on stabilization policies at the expense of development to claims of marginal gains in controlling price pressures."[16]

Criticism notwithstanding, the importance of World Bank and IMF involvement in African capacity innovation can be gleaned from a study by Sintim-Aboagye concerning the "relevance of the effect of economic programs of the World Bank and IMF on the economies of three African countries, Ghana, Senegal and Uganda to NEPAD's desired goal of creating an enabling environment with the aim to achieve sustained economic growth

---

[14] Benjamin F. Bobo and Hermann Sintim-Aboagye, op. cit. p. 269.
[15] See www.worldbank.org and www.imf.org.
[16] See Hermann Sintim-Aboagye, "IMF and World Bank Economic Programs on Inflation: Lessons for NEPAD," in Bobo and Sintim-Aboagye, op. cit., p. 147.

and development."[17] The study focused on the programs' effect on inflation and its uncertainty and employed a GARCH model to generate conditional variances as gauges for inflation uncertainty (inflation uncertainty is thought to cause inflation rates to either increase or decrease) and a Granger Causality procedure to test the direction of the link between inflation and inflation uncertainty.[18] Sintim-Aboagye reported that "Results of this study provide evidence suggesting that economic programs of the World Bank and IMF in Ghana, Senegal and Uganda contributed to stabilizing the macroeconomies of the three countries. The outcome shows that during the period of implementation of the programs both inflation and its uncertainty were lowered, ... Additionally, evidence also shows improvements in economic growth during the respective implementation periods of the economic programs."[19] With such empirical support, the Alliance is reassured that World Bank and IMF policies contribute to economic stabilization in African countries and to an enabling business environment conducive to capacity innovation strategy.

7.1.4 Non-Governmental Organizations

In the context of initiating development projects for the purpose of addressing grassroots issues and facilitating small-scale change, the NGO provides valuable input to capacity innovation. Grassroots activities invariably involve bottom up economic endeavors particularly when connected with projects such as small-scale farming, school programs, community water projects, parks and recreation, housing services, health care and the like.[20] Small-scale farming, for example, is particularly suited to capacity innovation. Where there is a desire by locals to engage in land cultivation, efforts by NEPAD to encourage such activity should be pursued given the importance of this matter as discussed in Chapter 4. The state typically focuses on large-scale projects and therefore unable to address community needs in a timely manner. The multinational lacks competitive advantage in undertaking small-scale community projects; their focus is on large-scale enterprise. And the NGO is not a source of FDI; their competitive advantage lies in arranging and

---

[17] Ibid., pp. 164–165.
[18] Ibid., pp. 150–151 and 271–280.
[19] Ibid., p. 165.
[20] Adrian Smith, Elisa Around, Mariano Fressoli, Hernan Thomas and Dinesh Abrol, "Supporting Grassroots Innovation: Facts and Figures," *Sci Dev Net*, http://www.scidev.net/global/icts/feature/supporting-grassroots-innovation-facts-and-figures-1.html, February 5, 2012.

providing assistance at the grassroots level. With NEPAD, multinationals and NGOs operating in a cross-fertilizing mode, efficient achievement of top down-bottom up capacity innovation is more assured.

Importantly, cross-fertilization particularly between multinationals and NGOs may take place through, as Bobo proposes, the "stakeholder givebacks" mechanism in which multinationals commit a portion of their profits to grassroots projects.[21] Stakeholder givebacks, when appropriately targeted, have the power to boost internal capacity. The NGO may play a strategic role here as intermediary; funds pass from MNC to NGO and from NGO to local projects for which the NGO provides technical assistance, management expertise and/or oversight.[22] This promotes Rostow's postulate on investment in capacity innovation in a dual fashion: FDI – with a twist, from top down and stakeholder givebacks from bottom up.[23]

Stakeholder givebacks, as Bobo confides, may raise concern about the profit maxim rule: mr = mc. Essentially, do stakeholder givebacks "disturb the precept that each factor of production is paid according to its worth and that the marginal productivity of each factor approximates to its price."[24] Bobo, however, purports no disturbance to the profit maxim rule by treating stakeholder givebacks essentially as profit sharing, a well-accepted medium of stakeholder (the employee) recognition. While there are only a few studies available on the relationship between profit sharing and profitability, empirical results generally obtain that profit sharing preserves the profit maxim. "FitzRoy and Kraft, using 1979 data on German metalworking firms, found positive effects of profit sharing on profitability. Using 1979–89 data on British firms, Bhargava also found a positive relationship between profit sharing and profitability. These findings were supported by Mitchell et al. using 1986 data on U.S. firms. Their results showed that profit sharing increases firm profits."[25]

---

[21] Bobo proposes "stakeholder givebacks" as a means of making FDI more user-friendly and in ways to boost internal capacity. See Bobo, op. cit., pp. 159–169 and 185–189.

[22] Many MNCs collaborate directly with established NGOs, which are well attuned to grassroots conditions. Andreas Wenger and Daniel Mockli, *Conflict prevention: The Untapped Potential of the Business Sector*, Boulder, Colo.: Lynne Rienner Publishers, 2003, p. 129.

[23] W.W. Rostow, *The Stages of Economic Growth: A Non-Communist Manifesto*, (Chapter 2, "The Five Stages of Growth—A Summary," pp. 4–16), Cambridge: Cambridge University Press, 1960.

[24] See Bobo, op. cit., p. 159.

[25] Ibid. See also Felix FitzRoy and Kornelius Kraft, "Profitability and Profit Sharing," *Journal of Industrial Economics*, 1986, 35, pp. 113–30; Sandeep Bhargava, 1994, "Profit Sharing and the Financial Performance of Companies: Evidence from U.K. Panel Data," *Economic Journal*, 104, pp. 1044–56; and Daniel J.B. Mitchell, David Lewin and Edward E. Lawler, III,

Hence, as Bobo asserts, stakeholder givebacks would be no less profit maxim preserving than profit sharing.[26]

Stakeholder givebacks aside, Amutabi presents a dissenting impression of NGOs in Africa, asserting that – being financed through proceeds of surplus capital (this may be likened to stakeholder givebacks), NGOs seek to dominate their constituents and other stakeholders and that their activities in agriculture, education and health sectors have not only facilitated development but undermined it as well – a sort of paradoxical duality suggesting that NGOs are agents of change while at the same time proxies of exploitation. Amutabi reckons NGO projects as opportunistic structures serving the process of globalization, more for the benefit of foreign interests than African needs.[27] Such opportunistic behavior, however, is reconciled by the opportunistic balancing structure presented in Chapter 6 in which the Africa-MNC Strategic Alliance minimizes any power imbalance thus curtailing exploitative motivations. An overarching function of the Alliance is to cast aside the status quo in restructuring the relationship between Africa and external interests purposely to orchestrate the most efficient means of arriving at sustained capacity innovation. All said, NEPAD's agenda is better served with both ends of the development spectrum engaged in activities that advance the capacity innovation objective.

7.1.5 Foreign Direct Investment

Formally the multinational's role begins with responding to NEPAD's expression of interest in foreign direct investment accompanied by rules of cooperation governing the diffusion of capital on the continent. Because of NEPAD's representation of African countries collectively, uniform policies and procedures regulating foreign capital may be more easily arranged and more transparent in application. Foreign ownership, competition, taxation and profits remittance typically require heady negotiation but NEPAD has the power to make this process less onerous. Doing so makes inherently good sense as Africa has such a tremendous need for foreign direct investment;

"Alternative Pay Systems, Firm Performance and Productivity," in Alan S. Blinder, ed., *Paying for Productivity: A Look at the Evidence*, Washington, DC: The Brookings Institute, pp. 15–95. Kim sheds further light on the relationship of profit sharing to profitability. See Seongsu Kim, "Does Profit Sharing Increase Firm's Profits?" *Journal of Labor Research*, 1998, 19(2), p. 351.

[26] Bobo, op. cit.

[27] See *JSTOR: Africa Today*, (review of *The NGO Factor in Africa: The Case of Arrested Development in Kenya*), Vol. 54, No. 4, Summer 2008, pp. 109–111; and Maurice N. Amutabi, *The NGO Factor in Africa: The Case of Arrested Development in Kenya*, New York: Routledge, 2006.

a population around one billion and GDP per capita below $2000 for half of the countries on the continent support this assertion.[28]

With terms of engagement in place, the multinational formally begins facilitation and expansion of capacity innovation effectively prompting growth of key sectors. Initiated by foreign direct investment and orchestrated through capital diffusion, technology transfer, new capital formation, and new capital formation spillover (positive), the scale of economic activity expands continuously, building economic capacity along the way. Capital diffusion spreads investment throughout the continent; technology transfer occasions transmission, absorption and assimilation of technology throughout the economy and workforce; new capital formation drives construction of production facilities as a direct result of capital diffusion, and new capital formation spillover produces indirect/side effects or associated benefits as derived from FDI. In short, FDI directly prompts development of key industries and spillover from FDI prompts emergence of support industries. For example, automobile manufacturing as a key industry gives rise to headlight and airbag manufacturing as support industries.

The multinational initiates production projects in sectors of its own choosing or in sectors targeted by NEPAD. However, as advised in Chapter 1, NEPAD must exercise caution here to avoid potential downsides of targeting. Critics of targeting argue that it is bad policy because market forces are better than politicians and bureaucrats at "picking winners."[29] Whether governments can pick winners more efficiently than markets remains a source of some debate; market proponents however see no contest.

7.1.6 Created Comparative Advantage

Apart from "natural" comparative advantage, developing "created" comparative advantage is crucial to capacity innovation and may be the single most important contribution to the process. Recall that "created" comparative advantage – a shortcoming of African countries, signals greater economic capacity than "natural" comparative advantage – a more prevalent feature on the continent. Brander instructs that "the orthodox position of international trade theory is that trade and investment patterns are determined by

---

[28] See Chapter 3, Exhibit 3.4.
[29] See Geoffrey Carliner, "Industrial Policies for Emerging Industries," in Paul R. Krugman, ed., *Strategic Trade Policy and the New International Economics*, Cambridge, Massachusetts: The MIT Press, 1990, p. 147.

comparative advantage, and that free markets are the best way of exploiting comparative advantage."[30] This matter was discussed in Chapter 3 but it is worth repeating here that comparative advantage is viewed as a matter of specialization in that a country has a comparative advantage in a product or service when it can produce that product or service at a lower cost than it can be produced by another country and, through trade among countries, everyone is better off because countries specialize in producing those products and services in which they have comparative advantage. Specialization, as argued, results in an efficient international division of labor whereby each country achieves a higher real national income and more robust market activity than it would have with no trade thus enabling it, in the context of African capacity innovation, to more aggressively achieve the objective.[31] Since theory underlying comparative advantage holds that world welfare – substitute here the African continent, is maximized when each country exports products whose comparative costs are lower at home than abroad and imports goods whose comparative costs are lower abroad than at home, engaging this concept is to Africa's advantage.[32] Importantly, the AU/NEPAD African Action Plan calls for "building productive capacity and capabilities for converting comparative advantage into industrial competitiveness."[33]

Comparative advantage is in many ways the driver of capacity innovation and the backbone of sustained development. A nation with the objective of capacity innovation is better equipped to achieve this objective based in part on its comparative advantage as reflected in relative factor endowment, that is – as Bertil Ohlin, the Swedish economist advises – the relative factor supplies of natural resources, labor and capital – the inputs necessary for production. Ohlin stressed the importance of relative factor supplies – noting that differences in relative factor supplies among countries produce differences in costs and prices, thereby forming a basis for international trade.[34] Technology and

---

[30] James A. Brander, "Rationales for Strategic Trade and Industrial Policy," in Paul R. Krugman, ed., *Strategic Trade Policy and the New International Economics*, Cambridge, Massachusetts: The MIT Press, 1990, p. 23.

[31] See Bobo, op. cit., pp. 3–4 and Gerald M. Meier, *Leading Issues in Economic Development*, New York: Oxford University Press, 1995, pp. 455–456.

[32] See Bobo, ibid., p. 99.

[33] See Revision of the AU/NEPAD AFRICAN ACTION PLAN 2010–2015: *Advancing Regional and Continental Integration Together through Shared Values*, Abridged Report 2010–2012, NEPAD Planning and Coordinating Agency, 2011, p. 42.

[34] See Meier, op. cit., p. 455.

technology factor differences also contribute to comparative advantage and, as Meier emphasizes, "nations have a comparative advantage in industries in which their firms gain a lead in technology."[35] An abundance of each factor is preferred; but developing countries tend to fall short, far short in many cases, on capital and technology. Since the capital and technology factors are well within the means of the multinational corporation, inviting in the multinational and getting on with the business of capacity innovation is a prudent decision. These very features are really what make the multinational corporation so important to African capacity innovation and the alliance arrangement so essential. The multinational's stock-in-trade, so to speak, is the facilitation of comparative advantage (as well as competitive advantage)[36] and since this is linked with capacity innovation, Africa's interests are served in alliance with the multinational. What better reason to align oneself with such capacity.

7.1.7 Regional Integration

Regional integration begins with national leaders in a region joining forces to connect their economies on matters of common interest, identifying compatible projects and harmonizing economic and investment policies and practices.[37] Regional integration as an Alliance objective is consistent with the World Bank study, *Africa's Infrastructure: A Time for Transformation*, noting that "With many small, isolated economies, Africa's economic geography is particularly challenging. Regional integration is likely the only way to overcome these handicaps and participate in the global economy.... It reduces costs, pools scarce technical and managerial capacity, and creates a larger market.... The goal of all regional . . . efforts is to facilitate the spatial organization of economic activity as a catalyst for faster growth. Lessons from the new economic geography, for which Paul Krugman received the Nobel Prize in Economics in 2008, explain this concept. Natural resources will

---

[35] Ibid., p. 456.

[36] It is argued in some circles that trade and gains from trade can arise independently of comparative advantage as multinationals exploit economies of scale and adopt strategies of product differentiation. See James R. Markusen, "The Boundaries of Multinational Enterprise and the Theory of International Trade," *Journal of Economic Perspectives*, Vol. 9, No. 2, Spring 1995, pp. 169–189.

[37] See Richard Llorah, "The Realities of Regional Integration: The NEPAD Perspective," in Bobo and Sintim-Aboagye, op. cit., pp. 171–172.

7.1 The Linking Process 153

remain important, but they provide few job opportunities and their benefits are seldom widely shared."[38]

Regional integration as an Alliance objective is also consistent with the African Union/NEPAD African Action Plan which prescribes that regional and continental integration across the full range of economic support activities are critical to development.[39] Though the extent to which the Plan will be accomplished any time soon is in question, the Africa-MNC Strategic Alliance proposed herein can play an important role. As the FDI and capital diffusion process unfolds under the purview of the Alliance, NEPAD promotes collective capacity innovation with particular attention to fostering regional integration in an effort to address Africa's critical mass problem. The multinational interfaces by spatially organizing production and supply of goods and services to support the effort. The critical mass problem of too few resources, too few people, too few skills, too few innovators and too little capital plagues African countries. Because many are mini-states, the problem in many instances is associated with smallness; domestic markets of mini-states are generally too small to accommodate enterprise of any efficient scale, thereby making it exceedingly difficult to sustain capacity innovation initiatives.[40] NEPAD has a mandate to promote self-reliance and through this to promulgate the matching of common interests among African countries and tying their different economies together to achieve economies of scale and develop regional markets;[41] a sort of 'of Africans and by Africans' initiative of self-reliance and economic integration. This involves complete harmonization of investment policies and full liberalization of trade in manufacturing and other sectors to substantially enhance trade volumes and prompt economic transformation. This magnitude of change is required to redress the critical mass problem. As part of the general policy apparatus, the multinational is looked upon as an initiator of change. Hence, with the multinational

---

[38] Vivien Foster and Cecilia Briceno-Garmendia, eds., *Africa's Infrastructure: A Time for Transformation*, Washington, DC: The International Bank for Reconstruction and Development/The World Bank, 2010, pp. 143–144.

[39] See *The AU/NEPAD African Action Plan 2010–2015: Advancing Regional and Continental Integration in Africa*, African Union, October 17, 2009, and *Revision of the AU/NEPAD AFRICAN ACTION PLAN 2010–2015: Advancing Regional and Continental Integration Together through Shared Values*, Abridged Report 2010–2012, NEPAD Planning and Coordinating Agency, 2011.

[40] See Bobo, op. cit., p. 141.

[41] See Richard Llorah, "The Realities of Regional Integration: The NEPAD Perspective," in Bobo and Sintim-Aboagye, op. cit., pp. 171–173.

playing the role of change agent, NEPAD has a legitimate tool with which to effect the necessary transformation. The multinational's characteristics give it tremendous capacity in this regard and the alliance relationship makes it a formidable ally.

It is noteworthy that the African Development Bank, the African Union Commission and the United Nations Economic Commission for Africa are collaborating on production of the Africa Regional Integration Index to track the progress of African countries and regional economic communities towards achieving regional integration goals. The Index will assist regional economic communities in identifying shortcomings and making informed policy decisions on corrective measures necessary to accomplish regional integration aspirations.[42] Findings from the 2016 Africa Regional Integration Index report indicate that while progress is being made, average integration scores stand at below half of the Index scale. It is such shortcomings that the Africa-MNC Strategic Alliance purports to address.

7.1.8 Economic Complementarity

Integrating African economies creates opportunity for addressing the problem of economic complementarity, a further weakness in African modernization capacity. With primary strength as suppliers of natural resources – a "natural" comparative advantage, African countries typically exhibit limited capacity to convert these resources to finished goods or demonstrate indifference toward integration schemes that purport to promote intra-continental complementarity.[43] And where there is finished goods output, the scale and quality are typically insufficient to achieve further comparative advantage. These incapacities limit trade among African countries as they are unable to supply mutual needs or to counterbalance mutual deficiencies; shortcomings that point to economic complementarity deficiency. The problem may be addressed to some extent if African countries produce and export goods and services among themselves related to their respective comparative advantage – those in which their absolute disadvantage is smaller, and import goods and services related to their respective comparative disadvantage – those in which their absolute disadvantage is greater. This process of production and

---

[42]*Africa Regional Integration Index, Report 2016*, African Union, African Development Bank and United Nations Economic Commission for Africa, 2016.

[43]See Robert S. Browne, "How Africa Can Prosper," in David N. Balaam and Michael Veseth, editors, *Readings in International Political Economy*, Upper Saddle River, New Jersey: Prentice-Hall, Inc., 1996.

trade guarantees efficient use of resources, increases output and promotes complementarity.[44] But the problem is not exclusively internal to the continent; international complementarity is a problem as well. Llorah remarks that "the income elasticity of demand for Africa's unprocessed foodstuffs and primary agricultural commodities is estimated at 0.6 percent and 0.5 percent, respectively, usually aggravated by the arbitrarily imposed so-called sanitary and phyto-sanitary rules that form extra layers of developed country regulatory barriers against African exports. These low elasticity estimates imply that African goods are considered inferior, and more so if the goods are fully manufactured ones."[45]

The inability to match the interests and requirements of one African country with a capacity to supply in another African country, and beyond, gives very strategic value to the Alliance. Working through the alliance arrangement, NEPAD and the multinational collectively promote and initiate production of goods and services to supply the mutual needs of African countries and for external markets through 'created' comparative advantage. In so doing, NEPAD encourages the substitution of machines for labor and moving toward higher technology production, of course keeping in mind the need for production that accommodates the high level of unskilled labor in Africa. Further, NEPAD promotes most favored-nation status and general preferential tariff among African countries to enhance complementarity. Concomitantly, the multinational pursues complex integration strategy – vertical and horizontal integration. In a three-country model of complex integration strategy and economic complementarity, Yeaple found that complex integration strategy creates complementarities between host countries.[46] In line with this, the multinational enhances complementarity through design and management of intra-continental supply chains. Supply chains encompass a network of interlinking activities and functions involved in extraction, manufacture and delivery of goods and services to consumers. This is particularly advantageous to African countries with comparative advantage in natural resources as supply chain operations have the capacity to extract and convert natural resources into end-user products and integrate this process into the African dynamic; a matter

---

[44] See Llorah, op. cit., p. 188.

[45] Ibid., p. 188.

[46] See Stephen Ross Yeaple, "The Complex Integration Strategies of Multinationals and Cross Country Dependencies in the Structure of Foreign Direct Investment," *Journal of International Economics*, Volume 60, Issue 2, August 2003, pp. 293–314. See also "Specialization, Complementarity and Multinational Industrialization in Africa: Progress Report," United Nations: Economic and Social Council; Economic Commission for Africa, 1975, Feb. 24–28.

of enhancing cross-country linkages and advancing economic integration.[47] These actions by NEPAD and the multinational potentially traverse a broad spectrum of complementarity: multi-way trade among African countries in goods, services and investments; people-to-people interaction as reflected in management and staff cross-border training programs; and government-to-government relationships associated with trade agreements and other means of cooperation. Complementarity has the benefit of addressing supply-side pressures among African countries as well as demand-side conditions; the pursuit of which prompts more robust capacity innovation.

While complementarity among African countries – combined with economic integration, may begin with comparative advantage in natural resources and labor-intensive goods, continuous activity through the alliance arrangement between NEPAD and the multinational and establishment of best practices across the capacity innovation framework will enable African countries to move beyond complementarity in low-cost sourcing to comparative advantage in capital intensive goods. This is the ultimate dynamic of capacity innovation. Matters of economic complementarity and economic integration will raise regulatory concerns to be sure, but NEPAD's mandate provides ample wherewithal to structure inter-country relationships necessary to regulate multi-way trade to promote efficiency, transparency and predictability in pursuing the capacity innovation framework.[48]

7.1.9 Price-Maker Status

As capital diffusion strategy gives rise to "created" comparative advantage, regional integration and economic complementarity, forces are set in place that spawn price-maker capacity within the collective facility of African countries. As they rely heavily upon exported natural resources and imported manufactured goods, the lucrative position of processed goods price making has

---

[47] Multinational supply chain operations offer a wide range benefits to African countries. See D.J. Ketchen, Jr. and G.T.M. Hult, "Building Theory About Supply Chain Management: Some Tools From The Organizational Sciencies," *Journal of Supply Chain Management*, Vol. 47, Issue 2, April 2011, pp. 12–18; Sunil Chapra and Peter Meindl, *Supply Chain Management*, 2nd ed. Upper Saddle River: Pearson Prentice Hall, 2004; and Arnaud Costinot, Jonathan Vogel and Su Wang, 2013, "An Elementary Theory of Global Supply Chains," *Review of Economic Studies*, Vol. 80, pp. 109–144.

[48] See "Canada-China Economic Complementarities Study," Economic Partnership Working Group, Foreign Affairs, Trade and Development Canada, Sept. 5, 2013, www.international.gc.ca./trade-agreements.

not been their strong suit. The limitation of "natural" comparative advantage restricts Africa's ability to substitute machines for labor and ultimately moving to higher technology production – necessary conditions for price-maker status. Price-making capacity stems from the notion that, as Browne affirms,[49] a country having established "created" comparative advantage places itself in a position to more efficiently compete with other countries in establishing the prices of goods and services; a matter of particular importance in international trade. The more competitive a country in this regard, the better able it is to influence prices – particularly prices of its goods and services. Neither generally able to competitively determine the prices of natural resource exports nor appreciably influence the prices of imported manufactured products, African countries are placed in a position of taking the prices they are offered rather than making the prices they merit. Relatively low prices for natural resources and high prices for manufactured imports – reflecting the propensity of international firms to extract maximum profit, make for an unhealthy functioning of the African system. Nourse in his notable work, *Price Making in a Democracy*, asserted very early on that "these situations call for recognition on the part of management that profits may at a given time be too high for the continued healthy functioning of the system."[50]

The U. S. Department of Agriculture, for example, addresses this matter from the perspective of small-scale producers – noting that "historically, farmers have been "price takers" when they deal with mainstream markets, ... – i.e., they exert little control over the prices of their goods." Recognizing that the production of end-user/consumer goods affords greater capacity to make/set prices, it is further noted that "... in direct-to-consumer marketing, farmers are "price makers," able to determine prices because their products have distinct characteristics, not easily substituted, that consumers want," and that "farmers can be price takers or price makers ... depending on how well they market themselves and their products' uniqueness/desirability and production capacity."[51]

Africa's shortcoming in this regard here again underscores the importance of an African capacity innovation strategy – in this instance that enables

---

[49] Browne, op. cit.

[50] Edwin G. Nourse, *Price Making in a Democracy*, Washington, D.C.: The Brookings Institution, 1944, p. 327.

[51] See Adam Diamond, James Barham and Debra Tropp, "Emerging Market Opportunities for Small-Scale Producers," Proceedings of a Special Session at the 2008 USDA Partners Meeting, U.S. Department of Agriculture, Agricultural Marketing Service, April, 2009, pp. 3 and 5, http://x.doi.org/10.9752/MS034.04-2009.

Africans to supply natural resources and produce manufactured products in ways that more efficiently facilitate control of prices. The Alliance puts in place a mechanism that promotes price maker outcomes.

7.1.10 Integrated Telecommunications

Connecting Africans and African commerce clearly are essential to capacity innovation on the continent. The mobile phone is doing much in the way of enabling Africans to keep pace with rapidly changing communications technology. Hosman and Fife confide that "the potential for mobile phones to reach a large and growing base of users across the continent, and to be used for development-related purposes, is becoming widely recognized, evidenced by the growing number of development-oriented projects, applications, and programs that specifically make use of mobiles."[52] But, as Hosman and Fife assert, "mobile phones remain an incomplete answer to the greater issue of underdevelopment...."[53] Importantly, a telecommunications superhighway complete with advanced information and communications technology (ICT) to integrate communications systems and connect people and communities throughout the continent and beyond is essential to long-term African capacity innovation. The AU/NEPAD African Action Plan prescribes that "access to advanced ICT is critical to the long-term economic and social development of Africa. It has increasingly become essential that appropriate ICT infrastructure, applications and skills are in place and accessible to the population to close the development gap between Africa and the rest of the world."[54] As proposed in the African-MNC Strategic Alliance, ICT would be aided through MNC capital diffusion in the form of direct investment in communications technology projects designed to support the continental goals of the Kigali Protocol: interconnect all African capitals and major cities and strengthen connectivity to the rest of the world; connect African villages to broadband ICT services and implement shared access initiatives such as community tele-centers and village phones; adopt key regulatory measures that promote affordable, widespread access to a full range of broadband ICT services; support the development of a critical mass of ICT skills

---

[52]Laura Hosman and Elizabeth Fife, "The Use of Mobile Phones for Development in Africa: Top-Down-Meets-Bottom-Up Partnering," *The Journal of Community Informatics*, Vol. 8, No. 3, 2012.

[53]Ibid.

[54]See *The AU/NEPAD African Action Plan 2010–2015: Advancing Regional and Continental Integration in Africa*, African Union, October 17, 2009, p. 25.

required by the knowledge economy; adopt a national e-strategy including a cyber-security framework, and deploy at least one flagship e-government service, e-education, e-commerce, and e-health service using accessible technologies in each country in Africa.[55] These goals are well within the skillset of multinationals and manageable through the Alliance. In fact, linking the continent to the world's information superhighway is underway and is the target of the undersea fiber optic cable initiative connecting African countries to Europe and Asia.[56] Broader participation by multinationals can speed up the process particularly in connecting intra-continental communications systems.

In the context of the Africa-MNC Strategic Alliance, functionally, NEPAD encourages FDI in telecommunications technology for the purpose of supporting the development of a telecommunications intra-continental superhighway. Communications technology multinationals respond to the call through initiation of communication projects that specifically link African communities throughout the continent via a network of communication technologies. Realization of the telecommunications superhighway is predicated upon development of a range of infrastructure to support the efficient diffusion of communications-related capital across the continent, particularly intra-continental roadways, power grids and Internet and broadband connectivity. The Alliance is especially equipped to more efficiently command resources for these purposes.

7.1.11 Infrastructure Development

The vital element in capacity innovation is clearly infrastructure development as it commands central priority in capacity innovation strategy. The entire structure of capacity innovation – spanning comparative advantage, regional economic integration and economic complementarity, relies on the infrastructure framework of a nation. Water, roads, electricity, ports, airports, and information and communications technology – all essential infrastructure components, facilitate the production and distribution of goods and services and promote a strong business environment. The World Bank study, *Africa's Infrastructure: A Time for Transformation*, sheds tremendous light on the state

---

[55]*The AU/NEPAD African Action Plan 2010–2015: Advancing Regional and Continental Integration in Africa*, ibid., p. 26.

[56]Marina Marais, *Technology: Linking Africa to the Information Superhighway*, http://www.ipsnews.net/2002/05/technology-linking-africa-to-the-information-superhighway/.

of infrastructure development across the continent (specifically Sub-Saharan Africa). Importantly, the study – covering 24 African countries, concludes that the cost of addressing Africa's infrastructure needs is an estimated $93 billion a year and that "deficiencies in infrastructure are holding back the continent by at least 1 percentage point in per capita growth" and further reports that "... infrastructure is a major constraint on doing business and depresses firm productivity by about 40 percent."[57] Inefficiencies in infrastructure operations and management cost Africa some $17 billion annually.[58] "By just about every measure of infrastructure coverage, African countries lag behind their peers in other parts of the developing world."[59] By comparison to South Asia, for example, with a similar per capita income, "In 1970, Africa had almost three times more electricity-generating capacity per million people than did South Asia. By 2000, South Asia had left Africa far behind – it now has almost twice the generating capacity per million people."[60] While the infrastructure problem is enormous, the cost of corrective measures is enormous as well. Based on country-level microeconomic modeling of Africa's infrastructure, "the cost to build new infrastructure, refurbish dilapidated assets, and operate and maintain all existing and new installations is estimated at almost $93 billion a year for 2006 through 2015."[61] Given the paucity in addressing these matters across the continent – "only three-fourths of the capital budgets allocated to infrastructure are actually executed," the cost of catch-up is likely much higher than current estimated need as the "estimates likely represent a conservative lower boundary for the cost of developing infrastructure assets at today's prices."[62]

Correcting infrastructure deficiency invariably points to financing. Not surprisingly, financing sources available to African countries have not provided sufficient funding to cover the estimated need; domestic public finance – the largest source, is over-subscribed. Increases in tax revenue and borrowing capacity show little promise; official development assistance (ODA) provided

---

[57] Vivien Foster and Cecilia Briceno-Garmendia, eds., *Africa's Infrastructure: A Time for Transformation*, Washington, DC: The International Bank for Reconstruction and Development/The World Bank, 2010, pp. 43 and 45; and Alvaro Escribano, J. Luis Guasch and Jorge Pena, "Impact of Infrastructure Constraints on Firm Productivity in Africa," *Working Paper 9*, Africa Infrastructure Sector Diagnostic, World Bank, Washington, DC, 2008.
[58] Vivien Foster and Cecilia Briceno-Garmendia, ibid., p. 67.
[59] Ibid., p. 47.
[60] Ibid., p. 48.
[61] Ibid., p. 58.
[62] Ibid., p. 65 and p. 60.

by the OECD has grown in recent years but only fills a portion of need. ODA to African infrastructure for most of the 1990s through early 2000s was rather constant at a trivial $2 billion a year. Finance by non-OECD countries, about the same as ODA, has been rising but future support is uncertain, and the financing is targeted primarily to oil-exporting countries; private participation is vulnerable to global market conditions. This capital – rising from about $3 billion in 1997 to well over $9 billion in 2007 and beyond, has been placed largely in resource-rich countries with water supply and sanitation attracting a meager share. Local capital markets have contributed on a relatively limited scale to infrastructure finance, with an outstanding stock of local infrastructure finance of around $15 billion. These markets tend to be relatively underdeveloped and constitute essentially commercial bank lending, some corporate bond and stock exchange activity, and a small number of institutional investors.[63] In light of these realities, the Africa-MNC Strategic Alliance is particularly important and especially in the context of private participation; the foreign direct investment focus and the regional integration tasking foretell promising outcomes for infrastructure finance. FDI-related infrastructure financing, particularly where new capital formation so warrants, holds interesting potential for private participation. Major corporate facilities require major support infrastructure, the financing of all or much of which conceivably rests with the corporation. Linking this effort with regional integration strategy not only bodes well for strengthening regional produce markets – as previously discussed, but also facilitates capital markets in the process. The World Bank study asserts that, regarding capital markets, "Regional integration of financial markets could achieve greater scale and liquidity. More cross-border intraregional listings – of both corporate bonds and equity issues, and more cross-border intraregional investment ... could help overcome national capital markets' impediments of small size, illiquidity, and inadequate market infrastructure. They could also facilitate the ability of companies and governments to raise financing for infrastructure. So far, this intraregional approach to raising infrastructure financing remains largely untapped."[64] Invariably the Study raises the question – what else can be done? By way of a reply, the Africa-MNC Strategic Alliance emphasizing private sector participation – in pursuing the measures of the African capacity innovation framework, conceivably holds very interesting promise for stimulating

---

[63] Ibid., pp. 75–81. The stock of local infrastructure finance is a projection based on the World Bank study and a rough estimate of increases since 2006.
[64] Ibid., p. 82.

activity that extends beyond current efforts in upgrading and modernizing African infrastructure.

A further question is raised by The World Bank study – does private sector participation work? Study findings are instructive. There are significant gains from private participation in some sectors, producing higher labor productivity – statistically significant in the cases of electricity and ports. There is also a close correlation between the competence in negotiating private participation deals and the efficiency gains by new private operatives. And, there is a correlation between institutional capacity and broader distribution of benefits; better institutional arrangements in deal making produce better outcomes for a wider range of stakeholders.[65] The study stresses that "institutional competence and capacity are important determinants of the performance of infrastructure providers in every sector" and especially that "The standard infrastructure reform and policy prescription package of the 1990s – market restructuring, private involvement up to and including privatization, establishing independent regulators, and enhancing competition, yielded a fair number of positive results in Africa. This conclusion deserves stress: beneficial outcomes following the application of these reforms have often been unacknowledged or at least underappreciated. Nevertheless, this set of reforms has proved more difficult to apply in Africa than in other regions. One finds in Africa numerous failures to implement, or fully implement, the policy package; renegotiations or cancellations of contracts with private providers; outcomes below expectations; and high degree of official and public skepticism about whether the application of the standard package is producing (or even could produce) the desired results. A large part of the explanation for this situation is thought to lie in the relative weakness of African practices, policies, and agencies (that is, institutions) that guide and oversee African infrastructure sectors and firms, public or private."[66]

Application of the noted reforms may have been particularly challenging owing to absence of an application framework from which institutional competence is more assured, best practices are more employed, and infrastructure activity is more efficient. The multinational is a prime example of extraordinary institutional competence (see Exhibit 7.2) and the Africa-MNC Strategic Alliance creates an institutional arrangement that provides even greater competence and capacity. NEPAD brings to the reform package a continental consensus of African leaders with change as a mandate – never

---

[65]Ibid., pp. 110–112.
[66]Ibid., p. 105.

before accomplished, and the multinational brings to the reform package the capacity of change agent – delivering a host of talent that is right for the job of capacity innovation. Working in concert as change agents – all players on the same page so to speak, engenders more efficient outcomes. This facilitates redress of the dire infrastructure situation in Africa, an undertaking that is a critical feature of the Alliance and the capacity innovation framework. Hence, private sector participation is workable and indeed offers efficient outcomes. The reform package – say the Africa-MNC Strategic Alliance: A Policy Framework for African Capacity Innovation, is a practical vehicle through which to engage reform, a credible platform upon which to stage transformation and a forward-looking business model capable of transcending inadequacies in the African marketplace.

7.1.12 NEPAD-MNC-NGO Convergence

Convergence of NEPAD, MNC and NGO interests as articulated in Chapter 4 particularly where complementary is a special benefit of the Alliance. The Alliance arrangement empowers NEPAD to coordinate and synchronize complementary activities thereby providing special inducement to the capacity innovation process that would not be the case otherwise. Particularly, linking MNC and NGO interests not just through stakeholder givebacks but through more direct interfacing such as the NGO responding to special needs of the multinational hastens capacity innovation in ways that each would not be able to otherwise accomplish, certainly not efficiently. As the multinational pursues its business plan, production requirements such as workers with special skill sets, for example, must be fulfilled. This need may be satisfied quite ably through NEPAD facilitation, MNC stakeholder givebacks and NGO assistance. The NGO for example, working with local school schemes and particularly technical institutes may facilitate preparation of prospective workers with required educational capacities and skill sets to respond to MNC demand and to other labor market demands as well. Through this mechanism, the NGO promotes technology transfer as well as job placement opportunities. This direct linking of MNC and NGO complementary interests occasions speedier capacity innovation down and up the development spectrum.

There has been, however, some deliberation concerning labor market intervention, particularly related to NGO initiatives. The World Bank offers the view that "The experience to date with labor market interventions in Africa is mixed, though a new generation of interventions is now being implemented. These range from the reform of formal vocational training systems to innovations in training and business development for young people. There are also some promising new approaches to promoting market-based skills development and providing complementary services that increase access to employment and income for the poor and vulnerable. Many of these approaches are focused on the informal sector, given its prominence in the African labor market. In the future, policymakers should seek to leverage scarce public resources by partnering with NGOs and the private sector to provide labor market services...."[67] Sanyal points out that that there are many arguments in support of NGOs but questions whether NGOs are particularly appropriate agents for fostering development from the bottom. In an empirical test of NGO capacity using results of field work in Bangladesh and India, Sanyal found that in the NGO's role in bottom-up projects, it is not effective in fostering development from below – asserting that "just as development does not trickle down from the top, neither does it effervesce from the bottom. Development requires a synergy between efforts made at the top and the bottom, a collaborative effort among the government, market institutions, and NGOs which utilizes the comparative advantage of each type of institution, and minimizes their comparative disadvantages."[68] In this context and to this end, the Africa-MNC alliance creates such a synergistic arrangement, focusing on the utility of comparative advantage and integration of comparative advantage in capacity innovation and the development process. In the Alliance, capacity innovation is viewed as collaboration among NEPAD (government), the multinational (market institution) and the NGO – NEPAD and the multinational having primary input and the NGO acting in a support but significant role.

7.1.13 Critical Mass

In ways critical mass is Africa's major hurdle; the large number of mini-states characterized by inefficient development and constrained by too little of all things necessary to infuse sustained transformation, from capital resources to human skills to supporting government structure. This matter has been reviewed in Chapter 3 and in the regional integration discourse above, but a few other insights are worthy of mention. Scott and Garofoli emphasize

---

[67]"Labor Market Programs," *Africa Social Protection Policy Briefs*, Washington, DC: The World Bank, p. 1.

[68]Bishwapriya Sanyal, *The Myth of Development from Below*, Massachusetts Institute of Technology, web.mit.edu/sanyal/www/articles/Myth%20of20Dev.pdf, pp. 1–3.

that more than a half-century ago Rosenstein-Rodan advocated critical mass as being important for successful development in the "big push" model – reasoning that a minimum investment in capacity innovation is necessary to overcome the obstacles to development.[69] From a geographical perspective of development, "regional push" effects, as Scott adds, are viewed as significant as well – inducing "virtuous circles of cumulative causation."[70] In the ordinary context, circular and cumulative causation is applied in a comparative mode to industrialized and developing countries, that is, the benefits derived from the continuous expansion and spread of multinational corporations are expected to accrue largely to the wealthy (industrialized countries) as opposed to those seeking to become wealthy (developing countries). But in the "virtuous" circular and cumulative causation context as applied to development of critical mass among countries across the African continent, – as proposed in the African capacity innovation framework herein, benefits are seen as accruing on moral and ethical grounds, that is, there is a reasonably fair sharing of benefits among African countries. Policy oversight through NEPAD's role in the Africa-MNC alliance promotes this outcome.

Further insight on critical mass is provided by Serfaty, informing that critical mass stands primarily for size – of country and population: in most instances bigger is better.[71] What this means for Africa is rather apparent. The fifty-five African economies include an abundance of economic units with populations considered too small to mount critical mass or to achieve an efficient scale of production – 38 below 20 million; 24 below 10 million;

---

[69] See Allen J. Scott and Gioacchino Garofoli, *Development on the Ground: Clusters, Networks and Regions in Emerging Economies*, New York: Routledge, 2007, p. 9; and P. Rosenstein-Rodan, "Problems of Industrialization of Eastern and South-Eastern Europe," *Economic Journal*, 53, pp. 202–211.

[70] Scott and Garofoli, ibid.; and A.J. Scott, "Regional Push: The Geography of Development and Growth in Low-and Middle-Income Countries," *Third World Quarterly*, 23, pp. 137–161. Absent policy interference, as proposed in the African capacity innovation framework herein, the process of circular and cumulative causation tends to reward the progressive and well-endowed region and even to thwart the efforts of the lagging region (see Gerald M. Meier, *Leading Issues in Economic Development*, New York: Oxford University Press, 1989, p. 385; and Benjamin F. Bobo, *Rich Country, Poor Country: The Multinational as Change Agent*, Westport, CT: Praeger Publishers, 2005, pp. 116–126). The notion of "virtuous circles of cumulative causation" imagines that with policy oversight such as prescribed by the African capacity innovation framework, development of critical mass takes place on moral and ethical grounds, that is, there is a reasonably fair sharing of critical mass benefits among African countries.

[71] Simon Serfaty, *A World Recast: An American Moment in a Post-Western Order*, New York: Rowman & Littlefield, 2012, p. 29.

and eight below 1.0 million, and too underdeveloped to create significant purchasing power parity, as seen through GDP per capita – 42 below $5000; 37 below $2000; and 10 below $1000.[72]

Meier also addresses the critical mass issue, noting that a "critical minimum effort" emphasizing the political and institutional framework and development of a market of sufficient scale to support a capital goods sector as being necessary to overcome the limitations of the small size of the domestic market.[73] And, Scott and Garofoli point out that "for a brief period in the 1970s, the notion of critical mass in development practice was downplayed by major international agencies in favor of so-called polarization-reversal policies. However, the policies failed dramatically precisely because networks of firms and their associated labor markets work most productively and innovatively where they achieve some minimal level of regional concentration."[74]

It is therefore noteworthy to emphasize that, ceteris paribus, the African capacity innovation framework herein prescribed accommodates: a minimum investment in capacity innovation through the foreign direct investment (multinational) approach to achieving critical mass; the notion that bigger is better through regional integration whereby African governments join forces to connect their economies on matters of common interest; and a "critical minimum effort" focused on political and institutional input and market development through employment of the Africa-MNC alliance strategy whereby NEPAD orchestrates the political and institutional machinery, the multinational propagates the market function and the NGO accommodates grassroots interests – the combined forces of which act to resolve the critical mass problem. As obtained through the Framework, a process of "virtuous" circular and cumulative causation is produced across the continent in which there is a reasonably fair sharing of critical mass benefits among African countries.

7.1.14 Transfer Pricing/Trade Misinvoicing

Transfer pricing – the pricing of goods and services traded among units of the same multinational corporation – has presented major problems for

---

[72] See Exhibit 3.4 in Chapter 3.

[73] Meier, op. cit., pp. 87 and 350; and Gerald M. Meier, *Leading Issues in Economic Development: Studies in International Poverty*, Second Edition, New York: Oxford University Press, 1970, p. 82.

[74] Scott and Garofoli, ibid.

African countries.[75] The difficulty essentially relates to the inability of African governments to investigate and monitor intra-firm pricing of commodity trade by multinationals, hence limiting their capacity to detect and control abuse. The abuse factor – manipulation of transfer prices or trade misinvoicing, derives from several benefits that may be realized: "the attainment or maintenance of market power or the penetration of new markets; lessening the impact of price controls; minimizing taxes and other payments to governments; and circumventing exchange controls and hedging against currency changes."[76] Leblanc maintains that Africa is losing billions of dollars annually through trade misinvoicing and that Global Financial Integrity estimates that $286 billion worth of capital was extracted out of Africa between 2002 and 2011 using this process.[77] Multinationals, however, "argue that established transfer prices are for the most part arm's length prices. In intra-firm transactions, the term arm's length is generally used to indicate the prices that would prevail if the two parties were unrelated. However, the term does not necessarily imply that such prices are established under competitive market conditions or that they are fair prices."[78] Importantly, where there is a difference between intra-firm and arm's length prices, African countries view this as a manipulation of transfer prices, hence the abuse factor.[79] But herein is the dilemma. Abuse

---

[75] The transfer pricing concept here broadly explicated is in instances a more complex notion particularly in the context of associated enterprises. See Ramon Dwarkasing, "Comments from Academia on the Revised Discussion Draft on Transfer Aspects of Intangibles," The Netherlands: Maastricht University, September 27, 2013, pp. 2–3.

[76] Dominant Positions of Market Power of Transnational Corporations: Use of the Transfer Pricing Mechanism, UNCTAD, 1978, p. 6. See also J. Greene and M. Duerr, *Intercompany Transactions in the Multinational Firm*, New York: The Conference Board, 1970, pp. 5–12, and Bobo, op. cit., p. 33.

[77] See Brian Leblanc, "Africa: Trade Misinvoicing, or How to Steal From Africa," allAfrica.com:Africa:Trade Misinvoicing, or How to Steal From Africa, May 6, 2014, p. 1. Global Financial Integrity is a Washington, DC-based think tank.

[78] Bobo, op. cit., p. 40, endnote 14. Dwarkasing advises that the notion of arm's length dealing is related to the doctrine of "undue influence." See Ramon Dwarkasing, ibid., p. 3. The African Tax Administration Forum provides extensive directives on transfer pricing policies and practices across the continent; the arm's length principle undergirds policies and practices pursued by African governments in multinationals' transactions. See "Transfer Pricing Updates Across Africa: EY Africa Tax Conference, September 2014," EY: Ernst & Young Global Limited, ey.com, 2014, pp. 1–36. See also Ramon Dwarkasing, Associated Enterprises: A Concept Essential for the Application of the Arm's Length Principle and Transfer Pricing, The Netherlands: Dwarkasing & Partners (Management @dwarkasing.com), 2011.

[79] Bobo, op. cit., p. 35.

is difficult to determine in light of "conceptual issues in defining appropriate transfer prices; an imbalanced or uneven incidence of transfer pricing across different industries and by different firms; internal and external problems gathering data relevant to checking and monitoring intra-firm pricing; and, management and procedural problems in monitoring transfer pricing and detecting improper conduct."[80]

Buehn and Eichler refer to trade misinvoicing as the dark side of world trade. This impression derives from their analysis of the determinants and development of trade misinvoicing using a data set of 86 countries over the period 1980 to 2005. They find that "trade misinvoicing occurs if the true value of exports or imports deviates from the amount of exports or imports entrepreneurs report to authorities" and that "trade misinvoicing is typically motivated by financial incentives, for example by benefiting from an existing premium at the black market of foreign exchange or by evading tariffs or taxes."[81] Patnaik et al. add further comment on the trade misinvoicing issue. They construct a 53-country data set over a 26 year span, covering industrialized and developing countries. Bifurcating trade misinvoicing into export misinvoicing and import misinvoicing, they find that capital account openness plays a major role in export misinvoicing but import misinvoicing is a result of two competing factors – "desire to keep capital out of the country leading to import overinvoicing, and the willingness to evade custom duties resulting in import underinvoicing."[82]

To address trade misinvoicing, what steps may be taken particularly regarding evasion of taxes on trade? Buehn and Eichler propose forming free trade agreements and reducing tariffs as remedy.[83] Patnaik et al. however question the tariffs reduction approach – commenting that when custom duties, i.e., tariffs, have been drastically reduced in developing countries ...why do they continue to play a vital role?[84] Kar and Freitas assert that

---

[80] Ibid., p. 32. Transfer pricing occurs in inter-firm trade as well. In this instance, under- or over-invoicing of openly traded goods has been widely observed in African countries. Ibid., p. 33.

[81] See Andreas Buehn and Stefan Eichler, "Trade Misinvoicing: the Dark Side of World Trade," Paper submitted to the Annual Meeting of the European Economic Association, Barcelona, 23–27 August 2009 (Conference Name: EEA 2009), www.Eea-esem.com/files/papers/EEA-ESEM/2009/1065/Buehn-Eichler.pdf, p. 1.

[82] Ila Patnaik, Abhijit Sen Gupta and Ajay Shah, "Determinants of Trade Misinvoicing," *Working Paper No. 2010-75*, New Delhi: National Institute of Public Finance and Policy, October 2010, http://www.nipfp.org.in.

[83] Buehn and Eichler, ibid., p. 22.

[84] Patnaik et al., op. cit., p. 22.

trade misinvoicing cannot be resolved by simply focusing on domestic policy measures that need to be taken by developing countries but that a host of policy measures is necessary to solve the problem of trade misinvoicing including "the requirement of country-by-country reporting by multinationals on their transactions and operations, and the automatic exchange of tax information between sovereign nations and tax havens."[85]

With this deliberation as backdrop, broadening the scope of trade misinvoicing remedy – not just to tackle trade misinvoicing singularly but the general issue of transfer pricing as well, warrants further reflection. While clearly trade misinvoicing takes center stage, every instance of transfer pricing is not clouded by misinvoicing. Even where transfer pricing involves arm's length prices, as multinationals argue are for the most part standard operating procedure, such pricing may not be sensitive to the economic condition of African countries, that is, even if or when prices are established under competitive market conditions, Africa's deep economic inadequacies beg for more lenient transfer pricing than perhaps the market mechanism obtains. Imagine, for example, the impact on capacity innovation of the unrealized $286 billion a year resulting from trade misinvoicing. Couple this with market pricing that results in multinationals paying lower taxes than may be the case otherwise. In the multinational structure where economic value added (the profit earned by the multinational less the cost of capital) may be more fully reflected in the final stage of product preparation – particularly in a multi-stage multi-country arrangement, profit maximizing strategy may yield extraordinary benefit to the corporation as desired profit may be more fully incorporated in end-user prices. In instances where all of these factors are at play, trade misinvoicing, market forces and profit maximizing behavior, African countries find themselves plagued by the Andre Gunder Frank proposition: "developing countries are little more than importers of goods and services from developed countries with price tags that far exceed their share of value added to the natural resource exports used in producing the imported goods and services."[86]

Even though multinationals purport fair pricing, African countries distrust the valuation process – certainly in the context of transfer pricing. Transfer pricing effectively impinges a range of concerns by African countries from abuse of trust to fair sharing of resource value to balance of power challenges – all of which are highlighted by trade misinvoicing but at the same time

---

[85] Dev Kar and Sarah Freitas, "Illicit Financial Flows from China and the Role of Trade Misinvoicing," *Global Financial Integrity*, October 2012, pp. 2 and 13.

[86] Bobo, op. cit., p. 14. See also Andre Gunder Frank, "The Development of Underdevelopment," *Monthly Review*, 1966, pp. 111–123.

impacted by the necessary procedure of transfer pricing. Hence to more fully address the transfer pricing issue, additional steps may prove beneficial, particularly in the context of the Africa-MNC Strategic Alliance.

First, the distrust issue in general must be reconciled. Bobo proposes the Global Interdependency Sensitivity Thesis (GIST) to address suspicion and distrust of the multinational as African countries view decisions taken by multinationals as products of a decidedly closed decision making process. GIST places Africa directly into the multinational's decision making process at the highest level of the corporation, the home based headquarters establishment.[87] In practice, implementing GIST involves "appointment to the corporate board of directors a contingent of government representatives from host countries who would participate in the conception of corporate policies and practices that affect the conduct of business in host developing nations and govern management decision-making at the subsidiary level in those countries ... to assure appropriate consideration of sensitive issues and concerns from both sides."[88] Under this arrangement, the mood of suspicion and distrust surrounding the Africa-MNC relationship would tend to diminish since the host country has a significant voice inside the corporation. This approach fits well in the Africa-MNC Strategic Alliance as the arrangement calls for intimate and direct interaction – a strong suit of the Alliance.

And, GIST has the additional capacity to impact fair sharing of resource value in that host representatives are afforded the opportunity to work directly with those setting transfer prices. "Evidencing its importance to MNCs, transfer-pricing decisions are usually made at the highest corporate levels. In fact, such decisions are accorded higher importance than market price decisions. The person responsible for setting transfer prices has a corporate rank of treasurer or higher. In most cases, however, this responsibility is assigned to the financial vice-president or comptroller. Interestingly, MNCs regard a decision on transfer pricing as the most critical short-run decision."[89] Transfer pricing practices under the GIST scheme become the purview of a more diverse decision-making body with definitive input from African representatives. Arm's length transfer pricing would tend not to invoke suspicion as the multinational's actions are for the most part transparent under GIST. The GIST arrangement does not vie to interdict multinational

---

[87] Bobo, op. cit., pp. 121–122.

[88] Ibid., p. 122.

[89] Ibid., p. 35. See also J.S. Arpan, *International Intracorporate Pricing: Non-American Systems and Views*, New York: Praeger, 1972, p. 111.

prerogative, but simply to improve relations with its hosts. Finally, balance of power challenges are common in the MNC-host country relationship and certainly in transfer pricing. GIST offers some response to this problem as it promotes shared decision making and through this a relative balancing of each party's inclination to take the opportunity to impose its will on the other. Beyond this, the opportunistic balancing structure proposed in the context of the Africa-MNC Strategic Alliance (see Chapter 6) has the wherewithal to bring further redress to the transfer pricing problem. The opportunistic balancing construct focuses directly on the power to impose one's will – in this instance by ensuring more transparent decision making and essentially more arm's length transfer pricing, hence easing pricing distortion; the aims being to encourage competitive pricing schemes and where possible to give special attention to Africa's economic condition. While under certain circumstances opportunistic behavior by multinationals in theory may be rational, that is, rent-seeking or self-interest seeking without guile, it may at the same time create undue hardship for African countries in practice. In the former context, Adam Smith in The Wealth of Nations wrote: "He [the economic agent] generally, indeed, neither intends to promote the public interest, nor knows how much he is promoting it. By preferring the support of domestic to that of foreign industry, he intends only his own security; and by directing that industry in such a manner as its produce may be of the greatest value, he intends only his own gain, and he is in this, as in many other cases, led by an invisible hand to promote an end which was no part of his intention. Nor is it always the worse for the society that it was no part of it. By pursuing his own interest he frequently promotes that of the society more effectually than when he really intends to promote it. I have never known much good done by those who affected to trade for the public good. It is an affectation, indeed, not very common among merchants, and very few words need be employed in dissuading them from it."[90] In the latter context, Al-Khatib et al. assert that opportunistic behavior can have a crippling effect on global business exchange in the form of decreased efficiency and inequitable distribution; the weaker exchange partner suffering inequity more punitively.[91] Hence, they suggest that opportunistic interaction by which two or more parties seek to do better through jointly-decided action mediates the tendency for each party

---

[90] Adam Smith, *The Wealth of Nations*, Franklin Center, Pennsylvania: The Franklin Library, 1978, p. 301.

[91] Jamal A. Al-Khatib, Avinash Malshe and John J. Sailors, "The Impact of Deceitful Tendencies, Relativism and Opportunism on Negotiation Tactics: A Comparative Study of US and Belgian Managers," *European Journal of Marketing, 45*(½), 2011, p. 135.

to maximize personal outcome at the expense of the other.[92] This notion is in direct harmony with the opportunistic balancing construct presented herein. NEPAD and the multinational as separate economic agents – with separate power, have the ability to behave opportunistically, however, in the confines of the Africa-MNC Strategic Alliance arrangement, each's behavior and power are bounded in the best interest of the Alliance, hence more balanced and less opportunistic transfer pricing decisionmaking.

7.1.15 Foreign Direct Investment Fund

The intent here is to further assist the provision of real resources – that is, beyond normal FDI activity, to support capacity innovation through increased capital accumulation and real investment in the capital stock. Foreign direct investment – the multinational corporation, does heavy lifting in this regard; FDI is an important factor in financing capital accumulation, as conferred heretofore. Apart from being important input to this book's dialogue, Bobo proposed it as the change agent in development in his book *Rich Country, Poor Country: The Multinational as Change Agent*, asserting that the deep poverty engulfing people of the Third World can be effectively addressed by the multinational corporation through capital diffusion and new capital formation.[93] Meier argues that more savings is needed to finance capital accumulation, proposing the release of resources from consumption – mobilization of an economic surplus, as means of increasing investment in real terms.[94] But what does this mean for Africans whose consumption level across the board, ergo economic surplus, is in many instances at bare minimum, so much so that the release of resources from consumption is a difficult feat no matter the source of consumption. Moreover, there are on the continent limited financial institutions to collect and channel savings into productive capacity innovation activities. This however is mindful of a study of the role of financial institutions particularly in cases of limited economic surplus: "However poor an economy may be there will be a need for institutions which allow such savings as are currently forthcoming to be invested conveniently and safely, and which ensure that they are channeled into the most useful purposes. The poorer a country is, in fact, the greater is the need for agencies to collect and invest the savings of the broad mass of persons and institutions within its borders. Such agencies will not only permit small amounts of savings to be handled and

---

[92] Ibid., p. 133.
[93] See Bobo, op. cit, pp. xiii–xiv.
[94] Meier, op. cit., p. 161.

invested conveniently but will allow the owners of savings to retain liquidity individually but finance long-term investment collectively."[95]

To be sure, there is a profound need on the continent for increased savings or savings-related resources and viable institutions to channel the resources to highest and best use. A supplement to "savings" – to the extent possible, that would essentially produce the same effect as real saving, that is, real investment in capital stock and an alternative mechanism to channel the resources to capacity innovation projects would serve Africa well. The multinational – through foreign direct investment, largely serves as a supplement to domestic savings and to some extent a substitute particularly early on in the transformation process.[96] As sustained transformation takes place, the input of "savings" relative to foreign direct investment would conceivably increase.

To this objective, the Foreign Direct Investment Fund (FDI Fund), as illustrated in Exhibit 7.3 further supplements "savings," assists capital accumulation and investment in capital stock, and provides a vehicle for funds distribution.[97] As foreign direct investment is a tremendous stimulant

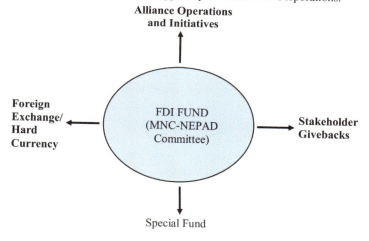

Exhibit 7.3 FDI Fund Support by Multinational Corporations.

---

[95] In Meier, ibid., p. 167. See also Edward Nevin, *Capital Funds in Underdeveloped Countries*, 1961, p. 75.

[96] Most developing countries supplement domestic savings with foreign aid, private foreign bank loans, private foreign investment and foreign direct investment.

[97] Bobo proposed the foreign direct investment fund in Benjamin F. Bobo, "Implementing State Interventionist Development: The Role of the Multinational Corporation," in Bobo and Sintim-Aboagye, op. cit., p. 219.

to capacity innovation activity, the multinational corporation obliges itself, if simply in the self-interest of image building and longevity, to support African nation building as a function of MNC largesse (that is, over and beyond new capital formation activities). This ancillary support provides critical financial input to the capacity innovation process. Organized under the structure of an FDI fund and administered by an MNC-NEPAD committee (perhaps housed in the Trust Department of a multinational bank) as portrayed in Exhibit 7.3, MNC ancillary support serves a number of purposes and takes place essentially through four media: Alliance Operations, Foreign Exchange/Hard Currency, Stakeholder Givebacks and Special Fund. Through these media, the Fund stands as an integrating factor in the Alliance, essentially functioning as a financing mechanism to (1) support Alliance administrative and operational activity – defraying the cost of implementing Alliance initiatives, (2) facilitate development of Alliance-related critical infrastructure – providing essential supporting amenities for new capital formation ordinarily the responsibility of government or that otherwise would not occur, particularly in a purely top down economic structure that would prevail without the "twist" as presented in Chapter 4, (3) support NGO programs – facilitating NGO activities through stakeholder givebacks particularly where these efforts directly serve the needs of the multinational, (4) facilitate "supply-leading" financial intermediary activity – for purposes of developing support industry and supply chains for multinational production,[98] (5) provide essential hard currency – for purposes of facilitating support industry and supply chains development, and supporting general Alliance initiatives, and (6) create a Special Fund (to be discussed in Chapter 8). In total – owing to the Alliance arrangement, the FDI Fund assists in producing a series of capacity innovation benefits that the continent would not experience otherwise.

Furthermore, referencing the currency issue, insufficient hard currency has an obviously pejorative impact on Africa's capacity innovation efforts. To address this matter, Africa would be well advised to adopt a triple-funds system – an intra-African currency, a foreign exchange reserve fund (outlined in Chapter 3) and a foreign direct investment fund (as articulated above). The intra-African currency is a single currency circulating only on the continent to

---

[98] Nafziger notes that financial intermediaries need not be limited to "demand-following" whereby they respond merely to investor and saver demand but may be "supply-leading" whereby they facilitate entrepreneurship and capital formation. See E. Wayne Nafziger, *The Economics of Developing Countries*, Second Edition, Englewood Cliffs, NJ: Prentice-Hall, Inc., 1990, p. 312.

facilitate regional integration and the exchange of goods and services among African countries. The foreign exchange reserve fund, perhaps linked to the Africa Trade Fund, facilitates global import/export activities pursuant to intra-continental needs – using foreign exchange origination outside the continent as a source of hard currency. And, the foreign direct investment fund – as outlined above, backstops the foreign exchange reserve fund, providing hard currency to address shortfalls.[99]

7.1.16 NEPAD Bank

To effect financial input (excluding retail banking) not undertaken by FDI, a commercial bank structured by NEPAD provides working capital and short-term loans to enterprises particularly serving the needs of multinationals and otherwise supporting NEPAD initiatives. As presented in Exhibit 7.4, the bank is capitalized by deposits from member countries, MNCs, the African Development Bank and local government. Operating fees will be provided by member countries and MNCs. Lending is the preferred activity but investing

Exhibit 7.4 NEPAD Bank High-Impact Loans/Investments Agricultural, Commercial, Industrial.

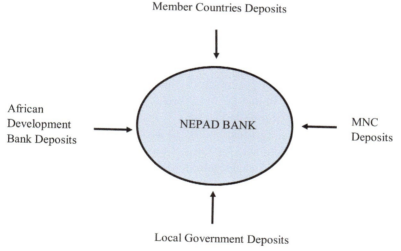

---

[99]For a broader presentation of the proposed currency system including the free-rider problem, see Benjamin F. Bobo, "Implementing State Interventionist Development: The Role of the Multinational Corporation," in Bobo and Sintim-Aboagye, op. cit., pp. 218–220.

176 *Path to Capacity Innovation: An Africa-MNC Strategic Alliance*

in essential projects such as infrastructure development may be required to encourage private initiative. Small-scale enterprises are also of interest to the Bank as they provide many products and services to multinational firms.

Member countries support the Bank through deposits and payment of fees to defray NEPAD's operating costs. Multinationals use the Bank as a depository for local operating and investment capital for corporate development and expansion; fees are paid to assist NEPAD with operating costs. The African Development Bank makes deposits in the Bank principally to support NEPAD; the Bank may in ways partner with the African Development Bank in co-financing arrangements, risk-mitigation instruments and direct capital diffusion initiatives.[100] Local government deposits public funds in the Bank to support NEPAD as well but, as in the case of all depositors, expects a return on the funds.

7.1.17 African Security Force-FDI Protection

Protecting the physical assets of multinational corporations is a formidable task and pursued at great financial and even human cost. The actual as well as expected attacks upon assets raise fears about FDI and discourage it altogether in many instances. Multinational corporations are often the targets of nefarious intentions giving rise to costly efforts to secure assets particularly vis-a-vis the physical security of plant, equipment, employees and corporate infrastructure. Rosenau, et al., investigators in a Rand Corporation study of multinational corporations as actors in conflict environments, note that MNCs pursue various means of protecting assets. They employed three case studies – Firestone in Liberia, Placer Dome in Papua New Guinea and Royal Dutch Shell in Nigeria, – as representative of the kinds of challenges MNCs face in zones of conflict and strategies pursued in responding to conflict situations. The least engaging response is providing safe havens for employees and their families during incidents of serious conflict; Firestone took this course of action at the height of the Liberian civil war.[101]

A more preemptive approach and common strategy, as Rand investigators suggest, is to undertake social investment and community development

---

[100] African Development Bank Group, http://www.afdb.org/en/about-us/mission-strategy/financing-the-strategy/.
[101] William Rosenau, Peter Chalk, Renny McPherson, Michelle Parker and Austin Long, "Corporations and Counterinsurgency," Intelligence Policy Center of the RAND National Security Research Division, Santa Monica, CA: Rand Corporation, 2009, pp. 2–3.

initiatives such as: building schools, hospitals and infrastructure and providing microcredit programs to support grassroots activities; engaging in "state-building" activities – sponsoring workshops to assist local authorities in improving their capacity to perform their duties and functions particularly related to peaceful coexistence; underwriting educational programs for police and military officials – an approach adopted by Placer Dome in Papua New Guinea;[102] and making direct investment to cleanup oil spills and providing loans to government to repair assets and restore oil production made necessary by militant attacks – an action taken by Royal Dutch Shell in Nigeria.[103]

These approaches to securing MNC assets in ways serve to win the hearts and minds of the populous and hopefully dissuade nefarious intentions; they do nothing however to directly protect plant and equipment. Moreover, there is a potential downside to engaging in community relations efforts – as the investigators warn – particularly should such activity inadvertently contribute to pejorative interethnic, religious, or tribal competition. This may result should the benefits flowing from an MNC operation – jobs, training, education, financial support, etc., favor one particular group over another.[104] Further, while these approaches essentially form a "best practices" scenario and suggest prudent means of mitigating and minimizing security threats, they may not be effective or appropriate for all circumstances. Hence, a more generally authoritative show of force or resistance capacity may be more practical. Chapter 8 takes up this matter in creating linkage and operating protocol between the Africa-MNC Strategic Alliance and the African Standby Force – a body overseen by the Peace and Security Council of the African Union. The African Standby Force has a range of responsibilities including confronting outright threats to or attacks upon persons or assets on the continent. NEPAD's mandate as conferred by the continent's leadership through the African Union provides the authority for this initiative as a key element in the linking process.

7.1.18 Country Capacity ID

Country capacity ID attributes are central to the promotion of modernity and progression of the socioeconomic system. The Africa-MNC Strategic Alliance seeks to employ the country capacity ID attributes in a manner most

---

[102]William Rosenau, et al, ibid., p. 4.
[103]Ed Crooks, "Shell to Lend Nigeria $3bn," *Financial Times*, FT.com, February 20, 2009.
[104]Rosenau, et al., ibid., p. 6.

advantageous to African capacity innovation. Bringing to fore and interfacing the components of the policy framework for African capacity innovation presented above initiate "upward movement of the entire social system"[105] and set the stage for attainment of the "ideals of modernization."[106] Over the long run, sustained progression of the socioeconomic system may be interpreted as successful capacity innovation.

7.1.19 Organizational Tie-in

The linking process requires direct organizational tie-in between NEPAD and the multinational. Bobo proposes a medium through which such a tie-in may be arranged in the Global Interdependency Sensitivity Thesis (GIST) discussed above. GIST proposes the organizational tie-in at the highest level of the corporation, where corporate policy decisions are made. In practice, as mentioned above, this involves appointment to corporate boards of directors NEPAD representatives who participate in the conception of corporate policies and practices that affect the conduct of the Alliance and the linking process (this procedure applies to the NEPAD board as well). In such arrangement, corporate and NEPAD policy makers work closely on an ongoing basis to assure appropriate engagement of the linking process and observance of Alliance objectives from both sides.[107]

In the NEPAD-multinational relationship, GIST bodes well for opportunistic balancing. In this context, trust in bargaining between host governments and multinationals is an abiding concern. A number of bargaining models purport to explain the negotiation process between host governments and multinationals. These models, as Chaitram Singh concludes, are essentially two forms: static bargaining models which underscore the initial negotiations between host government and the multinational, and dynamic models which embrace change over the life of the business arrangement.[108] "In the context of the static bargaining model, Robert Curry and Donald Rothchild propose a negotiation framework in which "impatience" and "reciprocal demand intensity" are the primary variables used to explain MNC-host government interactions. They argue that the host government's impatience to conclude business

---

[105] Myrdal, op. cit. and Meier, op. cit.
[106] Ibid.
[107] See Bobo, op. cit., pp. 111 and 121–122.
[108] See Chaitram Singh, *Multinationals, The State, and The Management of Economic Nationalism: The Case of Trinidad*, New York: Praeger, 1989, pp. 1–8; and Bobo, ibid., p. 119.

arrangements with the MNC works to its disbenefit, suggesting that the consequence of impatience is a contract that is unfavorable to the government. "...Curry and Rothchild's second variable, "reciprocal demand intensity" identifies how intensely the government desires what the multinational has to offer and vice versa. Typically, as they suggest, the government exhibits the greater demand intensity because of the absence of alternative firms with which to negotiate or strike a deal. The principal deficiency of the Curry-Rothchild model, as Singh points out, is that it largely ignores time in the business arrangement. Since the business arrangement is usually for an extended period of time, there exists the potential for renegotiation of the original agreement. Ignoring the issue of renegotiation does not allow consideration of the impact of the host country's domestic politics on the bargaining process. On this shortcoming of the Curry-Rothchild model, Singh suggests that an interactive bargaining model is needed to better [engage] the many facets of renegotiation."[109] To this end, the Global Interdependency Sensitivity Thesis was conceived to emphasize interactive bargaining between the host country and the multinational; raising the bargaining process to the highest level of the organization is a special feature of GIST. Ongoing interaction and decision making among top-level officials inherently benefit the bargaining relationship. GIST in combine with the Alliance mechanism directly mitigate the problems associated with "time," "reciprocal demand intensity" and "impatience." "Time" is addressed through the ongoing interaction among board members representing the multinational and NEPAD (host) and through the Alliance mechanism as an ongoing relationship; "Reciprocal demand intensity" is redressed through NEPAD's participation in corporate decision making and its relationship with a large number and variety of multinationals; and the "impatience" factor is minimized through high level corporate attention to the needs of NEPAD as well as through the Alliance's establishment of policies and practices that emphasize efficient contractual deal making. Board representation as provided by GIST and restriction of opportunistic behavior as emphasized by opportunistic balancing more assure an equitable and efficient organizational tie-in between NEPAD and the multinational. The tie-in is essentially the glue that binds the NEPAD-multinational relationship via the Africa-MNC Strategic Alliance and provides the defining dynamic for a holistic approach to the capacity innovation linking process.

---

[109] Robert L. Curry, Jr. and Donald Rothchild "On Economic Bargaining between African Governments and Multinational Companies," *The Journal of Modern African Studies*, 12, 2, 1974, 173–189; Chaitram Singh, ibid., pp. 5–6; and Bobo, op. cit., pp. 119–120.

7.1.20 Unlocking the Lock – Endgame

In Benjamin F. Bobo's book "Locked In and Locked Out," he explores the notion of the "locking effect": 'groups' being locked into economically distressed communities and locked out of more progressive environs. A particular rationale for this 'phenomenon' has important implication for the African predicament: "the locking effect perpetuates the system-maintenance goals of the dominant group...."[110] In the global economic order – in which Africa exists – the "dominant group" is seen through a variation of "hegemon theory" prescribing a global order in which dominant states set forth conditions of interaction backed up by powerful economies and military prowess. Attendantly, "dominant group" is also reflective of a variation of "power transition theory" in which "dominant states" – ones with the largest critical mass, GDP and political stability – headline a hierarchy of nations and control the allocation of global resources.[111]

In his exposition "The Development of Underdevelopment," Andre Gunder Frank argues that through foreign direct investment, the 'dominant group' – developed countries and their emissaries (multinational corporations) – have given rise to underdevelopment in Africa. This is buttressed by the assertion that the "developed countries usurp the strength of the developing countries by using them purely as exporters of the natural resources that are so important to fueling the engines of growth in developed countries ... and that developing countries are little more than importers of goods and services from developed countries with price tags that far exceed their share of value added to the natural resource exports used in producing the imported goods and services" thereby relegating developing countries to Third World status and effectively locking them into a position of underdevelopment.[112]

These edicts perhaps suggest that African countries may be more economically progressive by avoiding business relations with the 'dominant group'

---

[110] Benjamin F. Bobo, *Locked In and Locked Out: The Impact of Urban Land Use Policy and Market Forces on African Americans*, Westport, CT: Praeger Publishers, 2001, p. 6.

[111] See Mark Beavis, "The IR Theory Knowledge Base: IR Paradigms, Approaches and Theories," irtheory@hotmail.com, 1999–2014, pp. 9 and 16; Extract from lecture notes on the theory of hegemonic stability by Vincent Ferraro, Ruth C. Lawson Professor of International Politics, Mount Holyoke College, Massachusetts; and Ronald L. Tammen et al., *Power Transitions: Strategies for the 21st Century*, New York, NY: Seven Bridges Press, 2000.

[112] See Benjamin F. Bobo, *Rich Country, Poor Country: The Multinational as Change Agent*, Westport, Connecticut: Praeger Publishers, 2005, p. 14, and Andre Gunder Frank, "The Development of Underdevelopment," *Monthly Review*, 1966, pp. 111–123.

7.1 The Linking Process 181

particularly since 'rising underdevelopment' portends ominous impacts on capacity innovation strategy. This however would be ill-advised in light of the relatively weak economic position of African countries and their inability to compete with the power of the multinational corporation in global trade. As Bobo advises, "rather than rejecting the multinational's power, however valid the criticism, why not harness it? Why not use its great capacity to perform a wonder cure" for Africa's 'locked in' predicament?[113] Failure to unlock the lock and indeed, as Bobo asserts, "the inability to induce sustainable change may be due more to application of an ineffective change model than lack of effort."[114] The Africa-MNC Strategic Alliance capacity innovation model proposed herein offers such an opportunity as it harnesses the power of the multinational through an alliance relationship that minimizes dominant-group transgression through opportunistic balancing.

Exhibit 7.5 recapitulates Exhibit 7.1 in essence capturing the unlocking flow character of the core activities of the linking process as advanced in the capacity innovation framework. The flow process initializes with the multinational's decision to enter the African market in concert with NEPAD's decision to encourage MNC entry through institutional support and accommodation. These actions attract World Bank and IMF interest as they commit resources that support a nurturing environment in which the unlocking gestates. NGOs essentially round out organizational input through interfacing with the multinational, NEPAD, World Bank and IMF to promote especially grassroots linkage geared towards assuring a more holistic uplifting of the social strata. This organizational structure stimulates unlocking as the multinational implements capital diffusion across the continent – erecting new production facilities, impelled and backstopped by the supporting organizational units. As production takes hold, development of "created" comparative advantage ensues raising the necessity for a supportive market structure sufficient to achieve an efficient scale of production. With an abundance of economic units (countries) with populations considered too small to individually support an efficient scale of production – the critical mass problem, continued unlocking invokes regional integration of African markets. Markets integration sets the stage for development of economic complementarity among African countries – an unlocking of their incapacity to trade very significantly among themselves by way of supplying their mutual needs (goods and services) or

---
[113]Bobo, ibid., p. xxi.
[114]Ibid., p. xx.

Exhibit 7.5 Unlocking Core Activities of the Linking Process Africa—MNC Strategic Alliance.

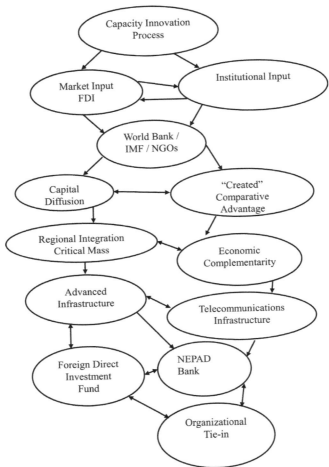

counterbalancing their mutual deficiencies (critical mass and 'created' comparative advantage). The range of elements – capital diffusion, comparative advantage, markets integration, economic complementarity, presupposes development of advanced infrastructure and particularly telecommunications to support the efficient production and distribution of goods and services. The unlocking experience is further induced by ancillary financial support orchestrated through the foreign direct investment fund and NEPAD bank – financing needs of capacity innovation, particularly occasioned by the

Alliance, not covered by FDI. And finally, sustained unlocking is precipitated through an organizational tie-in that provides basis for an enduring relationship between Africa and multinational corporations as the totality of the framework for African capacity innovation as orchestrated by the Africa – MNC Strategic Alliance occasions an endgame delivering capacity innovation dynamics that ultimately facilitate realistic near-future developed-country status for the African continent.

8

African Continental Security Apparatus: Accommodating FDI

Ultimately African capacity innovation efforts may be at risk of cooption by internal conflict that may arise through various means – war, insurrection, coup d'état, and the like. Unstable and insecure populations impact economic capacity. These raise the need for security policy to create and ensure a stable environment in which capacity innovation objectives may be pursued. Critical to successful capacity innovation will be creation of an intra-continental security apparatus to protect not only African citizens – as with the Protocol Relating to the Establishment of the Peace and Security Council of the African Union, and specifically the African Standby Force (ASF),[1] but also to ensure the integrity of the Africa-MNC alliance and thereby protect foreign direct investment. While protecting the citizenry is an obvious objective of security arrangements, should protecting FDI be considered a priority? Quite simply FDI is a prized commodity; multinational corporations are defined by it and home countries thrive on it. Western governments, ardent benefactors of FDI, are keen on its security, to say nothing of the expectations of multinationals. These assets contribute very significantly to profit margins and GNP.

But home countries lack the authority to exercise security measures as directly and forcefully in host countries as do FDI hosts. Given the high profile nature of FDI, host countries are expected to show an abiding concern for its security. As sovereign powers and masters of their security policy apparatus, hosts may intercede directly and by necessary means to interdict unlawful challenges to property and resources employed by foreign enterprises. Since FDI may play a central role in African capacity innovation strategies, conventional wisdom would suggest that African governments engage challenges to FDI as security objectives.

---

[1] See *Protocol Relating to the Establishment of the Peace and Security Council of the African Union* in the Appendix.

And, since FDI amounts to on-ground assets in African countries that form an economic arrangement in which they have a vested interest, security is quite literally obligatory. Protecting FDI is an investment in Africa's future. Success of the Africa-MNC alliance may well depend on it. Moreover, if the stages of economic growth, the ladder of comparative advantage, the core challenges and the capacity innovation endgame have relevance, a secure environment in which these processes and events materialize may give them a chance to do so.

To this end, an African security apparatus structured to accommodate FDI may be incorporated within the AU's peace and security architecture generally and within the Africa Standby Force mandate specifically. As a first move, expansion of Peace and Security Council protocol[2] is proposed as outlined in Exhibit 8.1. Importantly, the Council recognizes that sustainable development depends significantly upon a winning capacity innovation strategy which in turn is aided by Africa's ability to attract sustained foreign direct investment. But the African environment is viewed by many observers as a serious security risk. Recognizing the difficulty of attracting FDI under grave security risk, the Council expresses concern about the impact of threats and attacks against MNC assets as a matter of protocol. These pronouncements directly integrate the multinational corporation into security protocol; the MNC filling the joint role of capacity innovation facilitator and FDI input. In light of the importance of FDI to African development, the Council incorporates FDI security and protection throughout security protocol as demonstrated in Exhibit 8.1.

Further to the purpose of protecting FDI and in keeping with Peace and Security Council protocol, member countries agree to aid each other in response to threat or armed attack involving facilities or interests of multinational corporations. Essentially, the African Standby Force mandate offers an appropriate guide and template for structuring an actionable African security framework that accommodates foreign direct investment. Modified to reflect the FDI element, as outlined in Exhibit 8.2, the African security framework thereby presumes an intergovernmental security alliance of African countries agreeing to mutually defend and safeguard assets – capital and human – of multinational corporations. In the interest of a narrative for an African arrangement and employing the standing functions of the African Standby Force, the proposed functions for mandate to the African Standby Force regarding foreign direct investment security serve as a prospective model of an African FDI security scheme.

---

[2]Ibid.

> **Exhibit 8.1** Incorporating FDI Security.
>
> **Protocol Relating to the Establishment of the Peace and Security Council of the African Union**
>
> **FDI Protocol elements:** [See Protocol Relating to the Establishment of the Peace and Security Council of the African Union. Adopted by the 1st Ordinary Session of the Assembly of the African Union, Durban, 9 July, 2002.]
>
> - Aware further that sustainable development depends significantly upon African capacity innovation which in turn depends significantly upon foreign direct investment and capital diffusion on the continent.
> - Concerned about the impact of the threats and attacks against assets of multinational corporations and the difficulty of attracting FDI under grave security risk.
> - [Article 3.a.]: ...including presiding security and protection of foreign direct investment.
> - [Article 7.e.]: ...incorporate FDI as applicable.
> - [Article 7.e.]: "decide on any other issue having implications for the maintenance of peace, security, and stability on the continent," e.g., FDI ...
> - [Article 8.10.c.]: "any Regional Mechanism, international organization or civil society organization *including multinational corporations*...
> - [Article 12.3.]: ...academic institutions, NGOs *and multinational corporations*...
> - [Article 13.3.c.]: ...at the request of a Member State *or at the request of a multinational corporation as deemed appropriate by the AU*, ...
> - [Article 13.4.]: ...as well as national authorities, NGOs *and multinational corporations.*

The African Standby Force is currently charged with a range of peace support missions including observing and monitoring potential threats to African security. These responsibilities extend to more serious directive in the event of outright threat to or attack upon persons or assets. In such circumstance, the ASF is directed to intervene in order to restore peace and security and/or to prevent escalation or expansion of conflict, and ultimately to prevent conflict resurgence after settlement has been reached. After calm has been restored, the ASF takes on a peace-building mission, including post-conflict disarmament and demobilization. The ASF is further charged with humanitarian activity to alleviate the suffering of civilians exposed to conflict.

Importantly, the Peace and Security Council may mandate further functions to be carried out by the African Standby Force. This is particularly relevant here owing to the proposed functions for mandate to the ASF as delineated in Exhibit 8.2 regarding foreign direct investment. In conjunction with standing mandate, the ASF is charged with observing and monitoring potential threats to assets of multinational corporations. As proposed, in the event of outright threat to or attack upon such assets, the ASF intervenes to restore security and MNC control. Apart from ASF independent action, the

Exhibit 8.2 Foreign Direct Investment Security AU's Africa Peace and Security Architecture African Standby Force Mandate.

Functions mandated by Peace and Security Council Protocol to the African Standby Force: [See Protocol Relating to the Establishment of the Peace and Security Council of the African Union, Article 13, African Standby Force

a. Observation and monitoring missions;
b. Other types of peace support missions;
c. Intervention in a Member State in respect of grave circumstances or at the request of a Member State in order to restore peace and security, in accordance with Article 4(h) and (u) of the Constitutive Act;
d. Preventive deployment in order to prevent (i) a dispute or a conflict from escalating, (ii) an ongoing violent conflict from spreading to neighboring areas or States, and (iii) the resurgence of violence after parties to a conflict have reached an agreement
e. Peace-building, including post-conflict disarmament and demobilization;
f. Humanitarian assistance to alleviate the suffering of civilian population in conflict areas and support efforts to address major natural disasters; and
g. Any other functions as may be mandated by the Peace and Security Council or the Assembly.

Functions proposed for mandate by Peace and Security Council Protocol to the African Standby Force:

a. Observe and Monitor MNC Facilities
  - Ongoing surveillance deployment for large scale assets;
  - Periodic surveillance deployment for smaller scale assets;
b. Intervention in Member State to
  - Address grave circumstances regarding MNC facilities
  - Address the request of a Member State regarding MNC facilities,
c. Address the request of an MNC regarding MNC facilities, as deemed appropriate by UN body;
d. Preventive deployment in order to present
  - A dispute or conflict involving an MNC facility,
  - The resurgence of conflict after conflict resolution.
e. Peace-building, including post-conflict disarmament
  - Where the MNC has taken up arms to repel an attack
  - Where the attacking party has taken up arms against the MNC

Force also intervenes in conflict situations in response to requests by Member States regarding MNC facilities or to direct requests by MNCs. However, direct MNC requests are addressed as deemed appropriate by the AU body. Further, standing mandate on conflict prevention and peace-building extends to the ASF regarding multinational corporations. Here, the ASF deploys in order to prevent dispute or conflict involving an MNC facility and/or the resurgence of conflict after conflict resolution. In instances where arms have been employed during conflict involving an MNC, the ASF undertakes disarmament and peace-building to restore normality.

The proposed mandates send the message that FDI is regarded as an integral part of the African community and will be safeguarded as necessary to ensure its continued viability, so far as security is concerned. The message is particularly important to multinational corporations and to those so inclined as to commit assault upon MNC assets. The MNC is reassured that capital and human resources may be deployed safely and particularly in accordance with capital budgeting decision making; net present value will not be compromised by acts of indifference, disaffection, partisanship or in a broader realm – nationalistic fervor. Such acts may force the MNC into conflict with the host environment with unpredictable consequences. As for insurgent mindsets, the message is quite thought provoking. The African Standby Force is ready, willing and able to repel acts of aggression and especially attacks upon African assets, human or capital in nature. Multinational corporations, though of foreign origin, are accorded the same rights of protection and security as indigenous resources. Hence, MNCs are viewed as African assets with all the security rights and privileges thereto applied. In short, attack upon the MNC is attack upon Africa.

To provide adequate protection for African populations and resources, a strategic readiness response capability must be central to security protocol. Effectiveness of the African Standby Force will depend greatly upon the efficient positioning of ASF resources. Exhibit 8.3 purports strategic readiness response by the ASF as having a multitude of security force hubs located at key points on the continent, particularly population centers, schools, hospitals, main power plants, dams, ports, etc., and FDI facilities. The security capacity at each location may include troop transport vehicles, high tech communication devices, a minimum level of troops and support personnel and sufficient funding to ensure sustainability.

This arrangement is exemplified by Exhibit 8.3 in which major security bases already in place across the continent would be supported by security

Exhibit 8.3 Strategic Readiness Response Positioning of Security Bases African Standby Force.

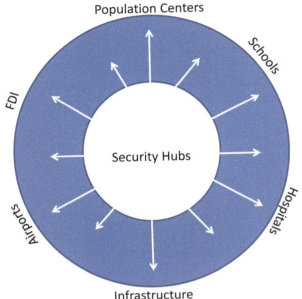

hubs and a cadre of smaller operational facilities (depicted by arrows) for quick response especially to smaller scale insurgencies. While such a strategic plan may appear excessively costly, the efficiency it offers in response time and protection of life and property has the effect of benefit outweighing cost. An analysis of strategic positioning of security bases by Wohlstetter suggests that locating bases far removed from targets may have potentially very significant cost implications; economizing on base locations will not necessarily mean economy in a total system of costs particularly when considering the costs of lives and property.[3]

Further, a comment by General Hoyt Vandenberg, former US Chief of Staff of the Northwest African Strategic Air Force,[4] is particularly informative here: "Obviously if you have to refuel two or three times in the air you get two or three missions per month out of your airplane instead of 15 or 20.

---

[3] A. J. Wholstetter, *Economic and Strategic Considerations in Air Base Location: A Preliminary Review*, Santa Monica, Calif.: RAND Corporation, D-1114, 29 December 1951.

[4] General Vandenberg was appointed chief of staff of the Northwest African Strategic Air Force on February 18, 1943. See http://www.af.mil/AboutUs/Biographies/Display/tabid/225/Article/105311/general-hoyt-s-vandenberg.aspx.

Therefore your Air Force, in order to do that efficiently, would have to be five or six times the size it would be if we had bases that were more nearly appropriate. Additionally it would take longer to have the full impact of the air war upon whoever we fought."[5] In this context, RAND proposes fewer bases but with regional hubs supported by a number of access sites (spokes) to maximize the ability to respond rapidly to small-scale threats and attacks.[6] The implication of these insights is instructive. Placing bases in a limited fashion across the vast African continent without hubs and spokes would be inefficient and unproductive. By the time a target is reached, insurgent forces would likely have completed their mission, established fortification or escaped.

Overseas base locations in U.S. military strategy offer even more insight on the topic. Today, the U.S. operates roughly 800 military bases in 63 countries with some 325,000 U.S. military personnel deployed in 156 countries at an annual cost of more than $100 billion. This effort provides a security system for the NATO alliance, Persian Gulf, Japan, South Korea, Taiwan and a host of others.[7]

Such resource allocation mapping and targets of security provide unquestionable strategic deterrence to potential adversaries. The moral to this story: if Africa truly desires internal security, it must take charge of providing it through its own initiative. A strategic system of bases, hubs and access sites sufficient to support efficient ready response capability is requisite to enable an African strategic deterrence to threats and attacks upon populations, infrastructure and FDI as well.

---

[5] Albert Wohlstetter and Harry Rowen, *Campaign Time Pattern, Sortie Rate, and Base Location*, Santa Monica, Calif.: RAND Corporation, P-1147, 25 January 1952. General Hoyt Vandenberg (four stars) served the US government and the US military in numerous capacities from 1923–1953 including director of Central Intelligence, chief of staff of the Air Force, commander of the Ninth Air Force, chief of staff of the 12th Air Force, and chief of staff of the Northwest African Strategic Air Force. See http://www.af.mil/AboutUs/Biographies/Display/tabid/225/Article/105311/general-hoyt-s-vandenberg.aspx.

[6] See Michael J. Lostumbo, et al., *Overseas Basing of U.S. Military Forces: An Assessment of Relative Costs and Strategic Benefits*, Santa Monica, Calif.: RAND Corporation, 2013. It should be noted that RAND found that, in light of annual recurring fixed costs, if forces were consolidated on fewer, larger bases, the fixed-cost portions of the closed bases would be saved. Further, as RAND confides, there are efficiencies to be gained from using fewer, larger bases rather than a more widely distributed system of bases. See Lostumbo, ibid.

[7] See *The Cost of U.S. Overseas Military Bases*, LessWaiting.com, April 23, 2012; http://www.globalsecurity.org/military/facility/images/diego-garcia-ims7.jpg; and Michael J. Lostumbo, et al., *Overseas Basing of U.S. Military Forces: An Assessment of Relative Costs and Strategic Benefits*, Santa Monica, Calif.: RAND Corporation, 2013.

192 African Continental Security Apparatus: Accommodating FDI

8.1 Special Fund

Ultimately, the African security system requires sustained financial support for ongoing implementation, maintenance and discharge of mandated responsibilities. The AU stipulates that the Peace and Security Council shall be supported by a Special Fund operated with contributions from member countries. In the interest of FDI security accommodation, Exhibit 8.4 proposes an arrangement in which multinational corporations share the financial obligations of the Special Fund. A component of the FDI Fund presented in Chapter 7, the Special Fund provision here provides an opportunity for MNCs to join African countries in carrying out AU security policy. An MNC presence in this regard does much for MNC image building and indeed the security of MNC assets.

In the specific context of African security, contributions to the FDI Fund by multinational corporations would be distributed, in part, to the Special Fund to directly support Peace and Security Council operations and especially the African Standby Force. From an overarching perspective, the security system virtually becomes the backbone of Africa's pursuit of a sustainable approach to capacity innovation and its drive to modernity in that it provides a secure and calm environment necessary for vibrant economic activity and sustained development. As the fourth component of the financial support structure, the

Exhibit 8.4 FDI Fund Special Fund Provision Peace and Security Council Support by Multinational Corporations.

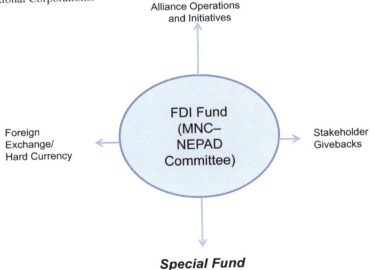

Special Fund essentially closes the loop on an organized system of ancillary but decidedly important financial input essential to assisting the funding of the full range of resource demands critically important to the capacity innovation process. The four-pronged structure forms an institution of sorts essentially underscoring institutional mapping discussed in Chapter 2 and brings a fitting conclusion to the capacity innovation narrative.

In light of the security policy apparatus adopted by the African Union, clearly consensus exists among African governments that security for the continent and a security pact representing the will of African nations are necessary and essential to the welfare of Africans. But, importantly, capacity innovation and in the long run sustained capacity innovation are keys to a meaningful Africa, a competitive Africa, a thriving Africa, and on and on. Sustained capacity innovation will quite literally unlock the door to sustained development and unleash the forces of change that will propel Africa into clear and indisputable modernity; the entirety of the foregoing pre-security dialogue, conceptual and prescriptive, has endeavored to so inform. But one further reality is inescapable; it has given the security discourse undeniable relevance. Be ever so advised that a secure environment is essential to the nourishment needed to seed and propagate capacity innovation. Therefore, African security protocol must necessarily provide accommodation for security of the very nourishment that is so essential to capacity innovation – foreign direct investment. Security narrative must profess the will and commitment to protect FDI – the multinational corporation – as an integral part of Africa itself. Sustained FDI will depend upon it, as will sustained capacity innovation.

9

Epilogue

For my entire career I have sought something that I have found in penning this book. This writing may be only of value to me. If so, the benefit of my effort walks in darkness. Certainly such is not my intent. The multitude of those toiling generally in the developing world and more specifically on the African continent is my motivation. They should not despair; the darkness is destined to give way to light. With its approach, my deepest optimism bears the gift of economic transformation as I hope to have transcended the presumed impossible. Underdevelopment is not a 'birthmark' characteristically an indelible brand but rather a 'lifemark' characteristically subject to the powers of change.

Jeremy Williams laments – why are some countries poor? – and counsels, often our answers seem to fall into two camps. "We either believe the rich countries exploit the poorer ones and it's the fault of the west, or we believe the poor countries are corrupt and pretty much deserve what they get. So we end up thinking of the third world as either helpless victims or money-grabbing scroungers. Neither of these extremes does justice to the third world and the predicament it is in."[1] Williams further submits that "In our modern world, wealth depends on trade, in goods or services. This is the essence of development. The more 'developed' a country is, the more money it is capable of making. So in talking about poor countries, . . . what we need to work out is why they are not developing. . . ."[2] I deeply and profoundly share this sentiment and hope that my effort in this work so attests.

The exigencies of development are often striking and even overwhelming but hopefully the prescriptive deliberation undertaken herein offers reason for optimism. The capacity innovation framework providing motive to the Africa-MNC Strategic Alliance is conceptualized with a view towards putting Africa on a course to sustained transformation and the expectation that it will

---

[1] Jeremy Williams, *Make Wealth History*, https://makewealthhistory.org/2007/11/05/why-are-some-countries-poor/

[2] Ibid.

indeed materialize. Capacity innovation is the key to Africa's transformation; with the appropriate catalysts, innovation and transformation are but a matter of time in gestation. The first of two major catalysts necessary to prompting this change so long sought by Africans came at the adoption of the New Partnership for Africa's Development. It is one of the most profound collaborations of African Heads of State. The second catalyst is proposed in this work in the form of the multinational corporation as change agent for the innovation process working in alliance with NEPAD as Africa's spokesperson for innovation. Important narrative and discourse undergirding the conceptualization of the process consume the pages of this work. The policy framework for African capacity innovation is the material product along with discourse for redress of corruption and security policy narrative for protecting the assets of multinational corporations. Arriving at these matters follows a deliberation structure that began with setting forth the direction of the work and identifying the objectives, then moving to dialogue purposely opening the door to concerns pertinent to understanding the African predicament and ultimately articulating a notion designed to remedy the conditions of suffering and distress emanating therefrom. Much is examined and liberties taken in conveying insights as deemed necessary.

Suggesting a more holistic approach to redressing Africa's capacity dilemma, this work focuses on the whole of the development spectrum and advocates an economic structural arrangement whereby private markets emerge and in such fashion as to promote efficiency in resource allocation resulting in market selection of economic drivers; questions capital theory as a development construct and an appropriate platform upon which sustained capacity innovation in Africa may emerge; explores Africa's road to modernity in the context of selected development constructs and assesses capacity innovation from a top down-bottom up perspective purposely to serve as backdrop to the Africa-MNC strategic alliance framework; constructs country capacity ID to identify internal resources available to African countries to support capacity innovation; conceptualizes the Africa-MNC strategic alliance to convey a capacity innovation philosophy; articulates an African capacity innovation policy framework to guide the Alliance through a series of actions designed to prompt innovation activity and set the continent on a course to sustained transformation; and articulates a scheme to protect assets – human and physical – derived through the Africa-MNC strategic alliance.

Running somewhat off grid with the conventional top down-bottom up development structure, a twist on the construct is advanced whereby the multinational corporation essentially enters the structure in alliance with the government to stimulate market forces from the top – spawning and bridging top-bottom convergence. Seen as speeding up the transformation process, the multinational serves to facilitate cross-fertilization between top and bottom, and among government, private sector and grassroots entities – resulting in prompting market forces from the bottom as well. Cross-fertilization enables emergence of a synergistic relationship between top and bottom not apparent heretofore.

The decidedly most worrying aspect of the alliance initiative – corruption – clearly exhibits the wherewithal to derail or contravene the capacity innovation process. It is important that the process not be subjugated to any adverse activity. Recall that under the African Union's principles, the Union has the right to intervene in a Member State when there are crimes against humanity. This work establishes that corruption is in fact a crime against humanity. The African Union therefore has the power to interdict corrupt practices, thus to protect the capacity innovation process.

As I bring this work to a close, hopefully it marks an ending and a beginning; ending to paltry economic transformation in Africa and, beginning to ascent upon a path to sustained modernity. Setting Africa upon this course, while indeed a rather complex undertaking, is attainable and well within our capacity to do so. Exhibit 9.1, the final exhibit in this work, reduces the matter

Exhibit 9.1 The Final Exhibit. The Africa-MNC Strategic Alliance Decision-Making Protocol.

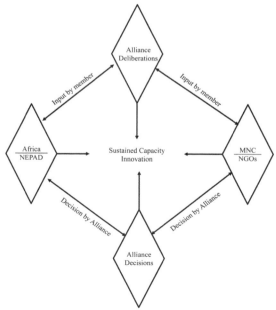

to a very simple common denominator: what's good for the Alliance is good for Africa; what's good for Africa is good for the Alliance. This is the essence of the Alliance pact and stands as the essence of the Alliance decision-making protocol.

The Alliance bespeaks a commitment by Africa to harness the powers of the multinational corporation in a relationship governed by a decision-making protocol. Ultimately engaging each other via the policy framework for African capacity innovation, the interaction plays out in expression of The Final Exhibit. Decision making undergirding the Alliance is enabled by input from Alliance members – NEPAD, MNCs and NGOs – processed through deliberations ultimately leading to decisions pertinent to the capacity innovation process. As Africa's voice in the Alliance, NEPAD ultimately has the weight of administration and success of mission on its shoulders; achieving sustained capacity innovation will be the judgement. The notion of engaging an alliance to form and administer a relationship is a universal concept. It forms the basis of trade agreements, military pacts, associations, and the like. It works in these instances, why not for Africa and multinationals.

Appendix

Protocol Relating to the Establishment of the Peace and Security Council of the African Union

WE, the Heads of State and Government of the Member States of the African Union;

CONSIDERING the Constitutive Act of the African Union and the Treaty establishing the African Economic Community, as well as the Charter of the United Nations;

RECALLING the Declaration on the establishment, within the Organization of African Unity (OAU), of a Mechanism for Conflict Prevention, Management and Resolution, adopted by the 29th Ordinary Session of the Assembly of Heads of State and Government of the OAU, held in Cairo, Egypt, from 28 to 30 June 1993;

RECALLING also Decision AHG/Dec.160 (XXXVII) adopted by the 3ih Ordinary Session of the Assembly of Heads of State and Government of the OAU, held in Lusaka, Zambia, from 9 to 11 July 2001, by which the Assembly decided to incorporate the Central Organ of the OAU Mechanism for Conflict Prevention, Management and Resolution as one of the organs of the Union, in accordance with Article 5(2) of the Constitutive Act of the African Union, and, in the regard, requested the Secretary-General to undertake a review of the structures, procedures and working methods of the Central Organ, including the possibility of changing its name;

MINDFUL of the provisions of the Charter of the United Nations, conferring on the Security Council primary responsibility for the maintenance of international peace and security, as well as the provisions of the Charter on the role of regional arrangements or agencies in the maintenance of international peace and security, and the need to forge closer cooperation and partnership between the United Nations, other international organizations and the African Union, in the promotion and maintenance of peace, security and stability in Africa;

ACKNOWLEDGING the contribution of African Regional Mechanisms for Conflict Prevention, Management and Resolution in the maintenance and promotion of peace, security and stability on the Continent and the need to develop formal coordination and cooperation arrangements between these Regional Mechanisms and the African Union;

RECALLING Decisions AHG/Dec.141 (XXXV) and AHG/Dec.142 (XXXV) on Unconstitutional Changes of Government, adopted by the 35th Ordinary Session of the Assembly of Heads of State and Government of the OAU held in Algiers, Algeria, from 12 to 14 July 1999, and Declaration AHG/Decl.S (XXXVI) on the Framework for an OAU Response to Unconstitutional Changes of Government, adopted by the 36th Ordinary Session of the Assembly of Heads of State and Government of the OAU, held in Lome, Togo, from 10 to 12 July 2000;

REAFFIRMING our commitment to Solemn Declaration AHG/Decl.4 (XXXVI) on the Conference on Security, Stability, Development and Cooperation in Africa (CSSDCA), adopted by the 36th Ordinary Session of the Assembly of Heads of State and Government of the OAU, held in Lome, Togo, from 10 to 12 July 2000, as well as Declaration AHG/Decl.1 (XXXVII) on the New Partnership for Africa's Development (NEPAD), which was adopted by the 37th Ordinary Session of the Assembly of Heads of State and Government of the OAU, held in Lusaka, Zambia, from 9 to 11 July 2001;

AFFIRMING our further commitment to Declaration AHG/Decl.2 (XXX) on the Code of Conduct for Inter-African Relations, adopted by the 30th Ordinary Session of the Assembly of Heads of State and Government of the OAU, held in Tunis, Tunisia, from 13 to 15 June 1994, as well as the Convention on the Prevention and Combating of Terrorism, adopted by the 35th Ordinary Session of the Assembly of Heads of State and Government of the OAU held in Algiers, Algeria, from 12 to 14 July 1999;

CONCERNED about the continued prevalence of armed conflicts in Africa and the fact that no single internal factor has contributed more to socio economic decline on the Continent and the suffering of the civilian population than the scourge of conflicts within and between our States;

CONCERNED ALSO by the fact that conflicts have forced millions of our people, including women and children, into a drifting life as refugees and internally displaced persons, deprived of their means of livelihood, human dignity and hope;

CONCERNED FURTHER about the scourge of landmines in the Continent and **RECALLING**, in this respect, the Plan of Action on a Landmine Free Africa, adopted by the 1st Continental Conference of African Experts on Anti-Personnel Mines, held in Kempton Park, South Africa, from 17 to 19 May 1997, and endorsed by the 66 1h Ordinary Session of the OAU Council of Ministers, held in Harare, Zimbabwe, from 26 to 30 May 1997, as well as subsequent decisions adopted by the OAU on this issue;

CONCERNED ALSO about the impact of the illicit proliferation, circulation and trafficking of small arms and light weapons in threatening peace and security in Africa and undermining efforts to improve the living standards of African peoples and **RECALLING**, in this respect, the Declaration on the Common African Position on the Illicit Proliferation, Circulation and Trafficking of Small Arms and Light Weapons, adopted by the OAU Ministerial Conference held in Bamako, Mali, from 30 November to 1 December 2000, as well as all subsequent OAU decisions on this issue;

AWARE that the problems caused by landmines and the illicit proliferation, circulation and trafficking of small arms and light weapons constitute a serious impediment to Africa's social and economic development, and that they can only be resolved within the framework of increased and well coordinated continental cooperation;

AWARE ALSO of the fact that the development of strong democratic institutions and culture, observance of human rights and the rule of law, as well as the implementation of post-conflict recovery programmes and sustainable development policies, are essential for the promotion of collective security, durable peace and stability, as well as for the prevention of conflicts;

DETERMINED to enhance our capacity to address the scourge of conflicts on the Continent and to ensure that Africa, through the African Union, plays a central role in bringing about peace, security and stability on the Continent;

DESIROUS of establishing an operational structure for the effective implementation of the decisions taken in the areas of conflict prevention, peace-making, peace support operations and intervention, as well as peace-building and post-conflict reconstruction, in accordance with the authority conferred in that regard by Article 5(2) of the Constitutive Act of the African Union;

Hereby Agree on the Following:

Article 1: Definitions

For the purpose of this Protocol:
  a. "Protocol" shall mean the present Protocol;
  b. "Cairo Declaration" shall mean the Declaration on the Establishment, within the OAU, of the Mechanism for Conflict Prevention, Management and Resolution;
  c. "Lome Declaration" shall mean the Declaration on the Framework for an OAU Response to Unconstitutional Changes of Government;
  d. "Constitutive Act" shall mean the Constitutive Act of the African Union;
  e. "Union" shall mean the African Union;
  f. "Assembly" shall mean the Assembly of Heads of State and Government of the African Union;
  g. "Commission" shall mean the Commission of the African Union;
  h. "Regional Mechanisms" shall mean the African Regional Mechanisms for Conflict Prevention, Management and Resolution;
  i. "Member States" shall mean Member States of the African Union.

Article 2: Establishment, Nature and Structure

1. There is hereby established, pursuant to Article 5(2) of the Constitutive Act, a Peace and Security Council within the Union, as a standing decision-making organ for the prevention, management and resolution of conflicts. The Peace and Security Council shall be a collective security and early-warning arrangement to facilitate timely and efficient response to conflict and crisis situations in Africa.

2. The Peace and Security Council shall be supported by the Commission, a Panel of the Wise, a Continental Early Warning System, an African Standby Force and a Special Fund.

Article 3: Objectives

The objectives for which the Peace and Security Council is established shall be to:

a. promote peace, security and stability in Africa, in order to guarantee the protection and preservation of life and property, the well-being of the African people and their environment, as well as the creation of conditions conducive to sustainable development;
b. anticipate and prevent conflicts. In circumstances where conflicts have occurred, the Peace and Security Council shall have the responsibility to undertake peace-making and peace-building functions for the resolution of these conflicts;
c. promote and implement peace-building and post-conflict reconstruction activities to consolidate peace and prevent the resurgence of violence;
d. co-ordinate and harmonize continental efforts in the prevention and combating of international terrorism in all its aspects;
e. develop a common defence policy for the Union, in accordance with Article 4(d) of the Constitutive Act;
f. promote and encourage democratic practices, good governance and the rule of law, protect human rights and fundamental freedoms, respect for the sanctity of human life and international humanitarian law, as part of efforts for preventing conflicts.

Article 4: Principles

The Peace and Security Council shall be guided by the principles enshrined in the Constitutive Act, the Charter of the United Nations and the Universal Declaration of Human Rights. It shall, in particular, be guided by the following principles:

a. peaceful settlement of disputes and conflicts;
b. early responses to contain crisis situations so as to prevent them from developing into full-blown conflicts;
c. respect for the rule of law, fundamental human rights and freedoms, the sanctity of human life and international humanitarian law;

d. interdependence between socio-economic development and the security of peoples and States;
e. respect for the sovereignty and territorial integrity of Member States;
f. non interference by any Member State in the internal affairs of another;
g. sovereign equality and interdependence of Member States;
h. inalienable right to independent existence;
i. respect of borders inherited on achievement of independence;
j. the right of the Union to intervene in a Member State pursuant to a decision of the Assembly in respect of grave circumstances, namely war crimes, genocide and crimes against humanity, in accordance with Article 4(h) of the Constitutive Act;
k. the right of Member States to request intervention from the Union in order to restore peace and security, in accordance with Article 4(j) of the Constitutive Act.

Article 5: Composition

1. The Peace and Security Council shall be composed of fifteen Members elected on the basis of equal rights, in the following manner:

a. ten Members elected for a term of two years; and
b. five Members elected for a term of three years in order to ensure continuity.

2. In electing the Members of the Peace and Security Council, the Assembly shall apply the principle of equitable regional representation and rotation, and the following criteria with regard to each prospective Member State:

a. commitment to uphold the principles of the Union;
b. contribution to the promotion and maintenance of peace and security in Africa – in this respect, experience in peace support operations would be an added advantage;
c. capacity and commitment to shoulder the responsibilities entailed in membership;
d. participation in conflict resolution, peace-making and peace-building at regional and continental levels;
e. willingness and ability to take up responsibility for regional and continental conflict resolution initiatives;
f. contribution to the Peace Fund and/or Special Fund created for specific purpose;

g. respect for constitutional governance, in accordance with the Lome Declaration, as well as the rule of law and human rights;
h. having sufficiently staffed and equipped Permanent Missions at the Headquarters of the Union and the United Nations, to be able to shoulder the responsibilities which go with the membership; and
i. commitment to honor financial obligations to the Union.

3. A retiring Member of the Peace and Security Council shall be eligible for immediate re-election.

4. There shall be a periodic review by the Assembly to assess the extent to which the Members of the Peace and Security Council continue to meet the requirements spelt out in Article 5(2) and to take action as appropriate.

Article 6: Functions

The Peace and Security Council shall perform functions in the following areas:
 a. promotion of peace, security and stability in Africa;
 b. early warning and preventive diplomacy;
 c. peace-making, including the use of good offices, mediation, conciliation and enquiry;
 d. peace support operations and intervention, pursuant to Article 4(h) and (j) of the Constitutive Act;
 e. peace-building and post-conflict reconstruction;
 f. humanitarian action and disaster management;
 g. any other function as may be decided by the Assembly.

Article 7: Powers

1. In conjunction with the Chairperson of the Commission, the Peace and Security Council shall:
 a. anticipate and prevent disputes and conflicts, as well as policies that may lead to genocide and crimes against humanity;
 b. undertake peace-making and peace-building functions to resolve conflicts where they have occurred;
 c. authorize the mounting and deployment of peace support missions;
 d. lay down general guidelines for the conduct of such missions, including the mandate thereof, and undertake periodic reviews of these guidelines;

e. recommend to the Assembly, pursuant to Article 4(h) of the Constitutive Act, intervention, on behalf of the Union, in a Member State in respect of grave circumstances, namely war crimes, genocide and crimes against humanity, as defined in relevant international conventions and instruments;
f. approve the modalities for intervention by the Union in a Member State, following a decision by the Assembly, pursuant to Article 4U) of the Constitutive Act;
g. institute sanctions whenever an unconstitutional change of Government takes place in a Member State, as provided for in the Lome Declaration;
h. implement the common defense policy of the Union;
i. ensure the implementation of the OAU Convention on the Prevention and Combating of Terrorism and other relevant international, continental and regional conventions and instruments and harmonize and coordinate efforts at regional and continental levels to combat international terrorism;
j. promote close harmonization, co-ordination and co-operation between Regional Mechanisms and the Union in the promotion and maintenance of peace, security and stability in Africa;
k. promote and develop a strong "partnership for peace and security" between the Union and the United Nations and its agencies, as well as with other relevant international organizations;
l. develop policies and action required to ensure that any external initiative in the field of peace and security on the continent takes place within the framework of the Union's objectives and priorities;
m. follow-up, within the framework of its conflict prevention responsibilities, the progress towards the promotion of democratic practices, good governance, the rule of law, protection of human rights and fundamental freedoms, respect for the sanctity of human life and international humanitarian law by Member States;
n. promote and encourage the implementation of OAU/AU, UN and other relevant international Conventions and Treaties on arms control and disarmament;
o. examine and take such appropriate action within its mandate in situations where the national independence and sovereignty of a Member State is threatened by acts of aggression, including by mercenaries;
p. support and facilitate humanitarian action in situations of armed conflicts or major natural disasters;

q. submit, through its Chairperson, regular reports to the Assembly on its activities and the state of peace and security in Africa; and

r. decide on any other issue having implications for the maintenance of peace, security and stability on the Continent and exercise powers that may be delegated to it by the Assembly, in accordance with Article 9 (2) of the Constitutive Act.

2. The Member States agree that in carrying out its duties under the present Protocol, the Peace and Security Council acts on their behalf.

3. The Member States agree to accept and implement the decisions of the Peace and Security Council, in accordance with the Constitutive Act.

4. The Member States shall extend full cooperation to, and facilitate action by the Peace and Security Council for the prevention, management and resolution of crises and conflicts, pursuant to the duties entrusted to it under the present Protocol.

Article 8: Procedure

Organization and Meetings

1. The Peace and Security Council shall be so organized as to be able to function continuously. For this purpose, each Member of the Peace and Security Council shall, at all times, be represented at the Headquarters of the Union.

2. The Peace and Security Council shall meet at the level of Permanent Representatives, Ministers or Heads of State and Government. It shall convene as often as required at the level of Permanent Representatives, but at least twice a month. The Ministers and the Heads of State and Government shall meet at least once a year, respectively.

3. The meetings of the Peace and Security Council shall be held at the Headquarters of the Union.

4. In the event a Member State invites the Peace and Security Council to meet in its country, provided that two-thirds of the Peace and Security Council members agree, that Member State shall defray the additional expenses

incurred by the Commission as a result of the meeting being held outside the Headquarters of the Union.

Subsidiary Bodies and Sub-Committees

5. The Peace and Security Council may establish such subsidiary bodies as it deems necessary for the performance of its functions. Such subsidiary bodies may include ad hoc committees for mediation, conciliation or enquiry, consisting of an individual State or group of States. The Peace and Security Council shall also seek such military, legal and other forms of expertise as it may require for the performance of its functions.

Chairmanship

6. The chair of the Peace and Security Council shall be held in turn by the Members of the Peace and Security Council in the alphabetical order of their names. Each Chairperson shall hold office for one calendar month.

Agenda

7. The provisional agenda of the Peace and Security Council shall be determined by the Chairperson of the Peace and Security Council on the basis of proposals submitted by the Chairperson of the Commission and Member States. The inclusion of any item in the provisional agenda may not be opposed by a Member State.

Quorum

8. The number of Members required to constitute a quorum shall be two-thirds of the total membership of the Peace and Security Council.

Conduct of Business

9. The Peace and Security Council shall hold closed meetings. Any Member of the Peace and Security Council which is party to a conflict or a situation under consideration by the Peace and Security Council shall not participate either in the discussion or in the decision making process relating to that conflict or situation. Such Member shall be invited to present its case to the Peace and Security Council as appropriate, and shall, thereafter, withdraw from the proceedings.

10. The Peace and Security Council may decide to hold open meetings. In this regard:
   a. any Member State which is not a Member of the Peace and Security Council, if it is party to a conflict or a situation under consideration by the Peace and Security Council, shall be invited to present its case as appropriate and shall participate, without the right to vote, in the discussion;
   b. any Member State which is not a Member of the Peace and Security Council may be invited to participate, without the right to vote, in the discussion of any question brought before the Peace and Security Council whenever that Member State considers that its interests are especially affected;
   c. any Regional Mechanism, international organization or civil society organization involved and/or interested in a conflict or a situation under consideration by the Peace and Security Council may be invited to participate, without the right to vote, in the discussion relating to that conflict or situation.

11. The Peace and Security Council may hold informal consultations with parties concerned by or interested in a conflict or a situation under its consideration, as well as with Regional Mechanisms, international organizations and civil society organizations as may be needed for the discharge of its responsibilities.

Voting

12. Each Member of the Peace and Security Council shall have one vote.

13. Decisions of the Peace and Security Council shall generally be guided by the principle of consensus. In cases where consensus cannot be reached, the Peace and Security Council shall adopt its decisions on procedural matters by a simple majority, while decisions on all other matters shall be made by a two-thirds majority vote of its Members voting.

Rules of Procedure

14. The Peace and Security Council shall submit its own rules of procedure, including on the convening of its meetings, the conduct of business, the publicity and records of meetings and any other relevant aspect of its work, for consideration and approval by the Assembly.

Article 9: Entry Points and Modalities for Action

1. The Peace and Security Council shall take initiatives and action it deems appropriate with regard to situations of potential conflict, as well as to those that have already developed into full-blown conflicts. The Peace and Security Council shall also take all measures that are required in order to prevent a conflict for which a settlement has already been reached from escalating.

2. To that end, the Peace and Security Council shall use its discretion to effect entry, whether through the collective intervention of the Council itself, or through its Chairperson and/or the Chairperson of the Commission, the Panel of the Wise, and/or in collaboration with the Regional Mechanisms.

Article 10: The Role of the Chairperson of the Commission

1. The Chairperson of the Commission shall, under the authority of the Peace and Security Council, and in consultation with all parties involved in a conflict, deploy efforts and take all initiatives deemed appropriate to prevent, manage and resolve conflicts.

2. To this end, the Chairperson of the Commission:
   a. shall bring to the attention of the Peace and Security Council any matter, which, in his/her opinion, may threaten peace, security and stability in the Continent;
   b. may bring to the attention of the Panel of the Wise any matter which, in his/her opinion, deserves their attention;
   c. may, at his/her own initiative or when so requested by the Peace and Security Council, use his/her good offices, either personally or through special envoys, special representatives, the Panel of the Wise or the Regional Mechanisms, to prevent potential conflicts, resolve actual conflicts and promote peace-building and post-conflict reconstruction.

3. The Chairperson of the Commission shall also:
   a. ensure the implementation and follow-up of the decisions of the Peace and Security Council, including mounting and deploying peace support missions authorized by the Peace and Security Council. In this respect, the Chairperson of the Commission shall keep the Peace and Security Council informed of developments relating to the functioning of such

missions. All problems likely to affect the continued and effective functioning of these missions shall be referred to the Peace and Security Council, for its consideration and appropriate action;
b. ensure the implementation and follow-up of the decisions taken by the Assembly in conformity with Article 4(h) and (j) of the Constitutive Act;
c. prepare comprehensive and periodic reports and documents, as required, to enable the Peace Security Council and its subsidiary bodies to perform their functions effectively.

4. In the exercise of his/her functions and powers, the Chairperson of the Commission shall be assisted by the Commissioner in charge of Peace and Security, who shall be responsible for the affairs of the Peace and Security Council. The Chairperson of the Commission shall rely on human and material resources available at the Commission, for servicing and providing support to the Peace and Security Council. In this regard, a Peace and Security Council Secretariat shall be established within the Directorate dealing with conflict prevention, management and resolution.

Article 11: Panel of the Wise

1. In order to support the efforts of the Peace and Security Council and those of the Chairperson of the Commission, particularly in the area of conflict prevention, a Panel of the Wise shall be established.

2. The Panel of the Wise shall be composed of five highly respected African personalities from various segments of society who have made outstanding contribution to the cause of peace, security and development on the continent. They shall be selected by the Chairperson of the Commission after consultation with the Member States concerned, on the basis of regional representation and appointed by the Assembly to serve for a period of three years.

3. The Panel of the Wise shall advise the Peace and Security Council and the Chairperson of the Commission on all issues pertaining to the promotion, and maintenance of peace, security and stability in Africa.

4. At the request of the Peace and Security Council or the Chairperson of the Commission, or at its own initiative, the Panel of the Wise shall undertake such action deemed appropriate to support the efforts of the Peace and Security Council and those of the Chairperson of the Commission for the prevention

of conflicts, and to pronounce itself on issues relating to the promotion and maintenance of peace, security and stability in Africa.

5. The Panel of the Wise shall report to the Peace and Security Council and, through the Peace and Security Council, to the Assembly.

6. The Panel of the Wise shall meet as may be required for the performance of its mandate. The Panel of the Wise shall normally hold its meetings at the Headquarters of the Union. In consultation with the Chairperson of the Commission, the Panel of the Wise may hold meetings at such places other than the Headquarters of the Union.

7. The modalities for the functioning of the Panel of the Wise shall be worked out by the Chairperson of the Commission and approved by the Peace and Security Council.

8. The allowances of members of the Panel of the Wise shall be determined by the Chairperson of the Commission in accordance with the Financial Rules and Regulations of the Union.

Article 12: Continental Early Warning System

1. In order to facilitate the anticipation and prevention of conflicts, a Continental Early Warning System to be known as the Early Warning System shall be established.

2. The Early Warning System shall consist of:
   a. an observation and monitoring centre, to be known as "The Situation Room", located at the Conflict Management Directorate of the Union, and responsible for data collection and analysis on the basis of an appropriate early warning indicators module; and
   b. observation and monitoring units of the Regional Mechanisms to be linked directly through appropriate means of communications to the Situation Room, and which shall collect and process data at their level and transmit the same to the Situation Room.

3. The Commission shall also collaborate with the United Nations, its agencies, other relevant international organizations, research centers, academic

institutions and NGOs, to facilitate the effective functioning of the Early Warning System.

4. The Early Warning System shall develop an early warning module based on clearly defined and accepted political, economic, social, military and humanitarian indicators, which shall be used to analyze developments within the continent and to recommend the best course of action.

5. The Chairperson of the Commission shall use the information gathered through the Early Warning System timeously to advise the Peace and Security Council on potential conflicts and threats to peace and security in Africa and recommend the best course of action. The Chairperson of the Commission shall also use this information for the execution of the responsibilities and functions entrusted to him/her under the present Protocol.

6. The Member States shall commit themselves to facilitate early action by the Peace and Security Council and or the Chairperson of the Commission based on early warning information.

7. The Chairperson of the Commission shall, in consultation with Member States, the Regional Mechanisms, the United Nations and other relevant institutions, work out the practical details for the establishment of the Early Warning System and take all the steps required for its effective functioning.

Article 13: African Standby Force
Composition

1. In order to enable the Peace and Security Council perform its responsibilities with respect to the deployment of peace support missions and intervention pursuant to Article 4(h) and (j) of the Constitutive Act, an African Standby Force shall be established. Such Force shall be composed of standby multidisciplinary contingents, with civilian and military components in their countries of origin and ready for rapid deployment at appropriate notice.

2. For that purpose, the Member States shall take steps to establish standby contingents for participation in peace support missions decided on by the Peace and Security Council or intervention authorized by the Assembly. The strength and types of such contingents, their degree of readiness and

general location shall be determined in accordance with established African Union Peace Support Standard Operating Procedures (SOPs), and shall be subject to periodic reviews depending on prevailing crisis and conflict situations.

Mandate

3. The African Standby Force shall, *inter alia,* perform functions in the following areas:
    a. observation and monitoring missions;
    b. other types of peace support missions;
    c. intervention in a Member State in respect of grave circumstances or at the request of a Member State in order to restore peace and security, in accordance with Article 4(h) and U) of the Constitutive Act;
    d. preventive deployment in order to prevent (i) a dispute or a conflict from escalating, (ii) an ongoing violent conflict from spreading to neighboring areas or States, and (iii) the resurgence of violence after parties to a conflict have reached an agreement.;
    e. peace-building, including post-conflict disarmament and demobilization;
    f. humanitarian assistance to alleviate the suffering of civilian population in conflict areas and support efforts to address major natural disasters; and
    g. any other functions as may be mandated by the Peace and Security Council or the Assembly.

4. In undertaking these functions, the African Standby Force shall, where appropriate, cooperate with the United Nations and its Agencies, other relevant international organizations and regional organizations, as well as with national authorities and NGOs.

5. The detailed tasks of the African Standby Force and its modus operandi for each authorized mission shall be considered and approved by the Peace and Security Council upon recommendation of the Commission.

Chain of Command

6. For each operation undertaken by the African Standby Force, the Chairperson of the Commission shall appoint a Special Representative and a Force Commander, whose detailed roles and functions shall be spelt out

in appropriate directives, in accordance with the Peace Support Standing Operating Procedures.

7. The Special Representative shall, through appropriate channels, report to the Chairperson of the Commission. The Force Commander shall report to the Special Representative. Contingent Commanders shall report to the Force Commander, while the civilian components shall report to the Special Representative.

Military Staff Committee

8. There shall be established a Military Staff Committee to advise and assist the Peace and Security Council in all questions relating to military and security requirements for the promotion and maintenance of peace and security in Africa.

9. The Military Staff Committee shall be composed of Senior Military Officers of the Members of the Peace and Security Council. Any Member State not represented on the Military Staff Committee may be invited by the Committee to participate in its deliberations when it is so required for the efficient discharge of the Committee's responsibilities.

10. The Military Staff Committee shall meet as often as required to deliberate on matters referred to it by the Peace and Security Council.

11. The Military Staff Committee may also meet at the level of the Chief of Defence Staff of the Members of the Peace and Security Council to discuss questions relating to the military and security requirements for the promotion and maintenance of peace and security in Africa. The Chiefs of Defence Staff shall submit to the Chairperson of the Commission recommendations on how to enhance Africa's peace support capacities.

12. The Chairperson of the Commission shall take all appropriate steps for the convening of and follow-up of the meetings of the Chiefs of Defence Staff of Members of the Peace and Security Council.

Training

13. The Commission shall provide guidelines for the training of the civilian and military personnel of national standby contingents at both operational and

tactical levels. Training on International Humanitarian Law and International Human Rights Law, with particular emphasis on the rights of women and children, shall be an integral part of the training of such personnel.

14. To that end, the Commission shall expedite the development and circulation of appropriate Standing Operating Procedures to *inter-alia:*
   a. support standardization of training doctrines, manuals and programmes for national and regional schools of excellence;
   b. co-ordinate the African Standby Force training courses, command and staff exercises, as well as field training exercises.

15. The Commission shall, in collaboration with the United Nations, undertake periodic assessment of African peace support capacities.

16. The Commission shall, in consultation with the United Nations Secretariat, assist in the co-ordination of external initiatives in support of the African Standby Force capacity-building in training, logistics, equipment, communications and funding.

Role of Member States

17. In addition to their responsibilities as stipulated under the present Protocol:
   a. troop contributing countries States shall immediately, upon request by the Commission, following an authorization by the Peace and Security Council or the Assembly, release the standby contingents with the necessary equipment for the operations envisaged under Article 9 (3) of the present Protocol;
   b. Member States shall commit themselves to make available to the Union all forms of assistance and support required for the promotion and maintenance of peace, security and stability on the Continent, including rights of passage through their territories.

Article 14: Peace Building

Institutional Capacity for Peace-building

1. In post-conflict situations, the Peace and Security Council shall assist in the restoration of the rule of law, establishment and development of democratic

institutions and the preparation, organization and supervision of elections in the concerned Member State.

Peace-building During Hostilities

2. In areas of relative peace, priority shall be accorded to the implementation of policy designed to reduce degradation of social and economic conditions arising from conflicts.

Peace-building at the End of Hostilities

3. To assist Member States that have been adversely affected by violent conflicts, the Peace and Security Council shall undertake the following activities:
   a. consolidation of the peace agreements that have been negotiated;
   b. establishment of conditions of political, social and economic reconstruction of the society and Government institutions;
   c. implementation of disarmament, demobilization and reintegration programmes, including those for child soldiers;
   d. resettlement and reintegration of refugees and internally displaced persons;
   e. assistance to vulnerable persons, including children, the elderly, women and other traumatized groups in the society.

Article 15: Humanitarian Action

1. The Peace and Security Council shall take active part in coordinating and conducting humanitarian action in order to restore life to normalcy in the event of conflicts or natural disasters.

2. In this regard, the Peace and Security Council shall develop its own capacity to efficiently undertake humanitarian action.

3. The African Standby Force shall be adequately equipped to undertake humanitarian activities in their mission areas under the control of the Chairperson of the Commission.

4. The African Standby Force shall facilitate the activities of the humanitarian agencies in the mission areas.

Article 16: Relationship with Regional Mechanisms for Conflict Prevention, Management and Resolution

1. The Regional Mechanisms are part of the overall security architecture of the Union, which has the primary responsibility for promoting peace, security and stability in Africa. In this respect, the Peace and Security Council and the Chairperson of the Commission, shall:
   a. harmonize and coordinate the activities of Regional Mechanisms in the field of peace, security and stability to ensure that these activities are consistent with the objectives and principles of the Union;
   b. work closely with Regional Mechanisms, to ensure effective partnership between them and the Peace and Security Council in the promotion and maintenance of peace, security and stability. The modalities of such partnership shall be determined by the comparative advantage of each and the prevailing circumstances.

2. The Peace and Security Council shall, in consultation with Regional Mechanisms, promote initiatives aimed at anticipating and preventing conflicts and, in circumstances where conflicts have occurred, peace-making and peace-building functions.

3. In undertaking these efforts, Regional Mechanisms concerned shall, through the Chairperson of the Commission, keep the Peace and Security Council fully and continuously informed of their activities and ensure that these activities are closely harmonized and coordinated with the activities of Peace and Security Council. The Peace and Security Council shall, through the Chairperson of the Commission, also keep the Regional Mechanisms fully and continuously informed of its activities.

4. In order to ensure close harmonization and coordination and facilitate regular exchange of information, the Chairperson of the Commission shall convene periodic meetings, but at least once a year, with the Chief Executives and/or the officials in charge of peace and security within the Regional Mechanisms.

5. The Chairperson of the Commission shall take the necessary measures, where appropriate, to ensure the full involvement of Regional Mechanisms in the establishment and effective functioning of the Early Warning System and the African Standby Force.

6. Regional Mechanisms shall be invited to participate in the discussion of any question brought before the Peace and Security Council whenever that question is being addressed by a Regional Mechanism is of special interest to that Organization.

7. The Chairperson of the Commission shall be invited to participate in meetings and deliberations of Regional Mechanisms.

8. In order to strengthen coordination and cooperation, the Commission shall establish liaison offices to the Regional Mechanisms. The Regional Mechanisms shall be encouraged to establish liaison offices to the Commission.

9. On the basis of the *above* provisions, a Memorandum of Understanding on Cooperation shall be concluded between the Commission and the Regional Mechanisms.

Article 17: Relationship with the United Nations and Other International Organizations

1. In the fulfillment of its mandate in the promotion and maintenance of peace, security and stability in Africa, the Peace and Security Council shall cooperate and work closely with the United Nations Security Council, which has the primary responsibility for the maintenance of international peace and security. The Peace and Security Council shall also cooperate and work closely with other relevant UN Agencies in the promotion of peace, security and stability in Africa.

2. Where necessary, recourse will be made to the United Nations to provide the necessary financial, logistical and military support for the African Unions' activities in the promotion and maintenance of peace, security and stability in Africa, in keeping with the provisions of Chapter VIII of the UN Charter on the role of Regional Organizations in the maintenance of international peace and security.

3. The Peace and Security Council and the Chairperson of the Commission shall maintain close and continued interaction with the United Nations Security Council, its African members, as well as with the Secretary-General,

including holding periodic meetings and regular consultations on questions of peace, security and stability in Africa.

4. The Peace and Security Council shall also cooperate and work closely with other relevant international organizations on issues of peace, security and stability in Africa. Such organizations may be invited to address the Peace and Security Council on issues of common interest, if the latter considers that the efficient discharge of its responsibilities does so require.

Article 18: Relationship with the Pan African Parliament

1. The Mechanism shall maintain close working relations with the Pan-African Parliament in furtherance of peace, security and stability in Africa.

2. The Peace and Security Council shall, whenever so requested by the Pan African Parliament, submit, through the Chairperson of the Commission, reports to the Pan-African Parliament, in order to facilitate the discharge by the latter of its responsibilities relating to the maintenance of peace, security and stability in Africa.

3. The Chairperson of the Commission shall present to the Pan-African Parliament an annual report on the state of peace and security in the continent. The Chairperson of the Commission shall also take all steps required to facilitate the exercise by the Pan-African Parliament of its powers, as stipulated in Article 11(5) of the Protocol to the Treaty establishing the African Economic Community relating to the Pan-African Parliament, as well as in Article 11(9) in so far as it relates to the objective of promoting peace, security and stability as spelt out in Article 3(5) of the said Protocol.

Article 19: Relationship with the African Commission on Human and Peoples' Rights

The Peace and Security Council shall seek close cooperation with the African Commission on Human and Peoples' Rights in all matters relevant to its objectives and mandate. The Commission on Human and Peoples' Rights shall bring to the attention of the Peace and Security Council any information relevant to the objectives and mandate of the Peace and Security Council.

Article 20: Relations with Civil Society Organizations

The Peace and Security Council shall encourage non-governmental organizations, community-based and other civil society organizations, particularly women's organizations, to participate actively in the efforts aimed at promoting peace, security and stability in Africa. When required, such organizations may be invited to address the Peace and Security Council.

Article 21: Funding

Peace Fund

1. In order to provide the necessary financial resources for peace support missions and other operational activities related to peace and security, a Special Fund, to be known as the Peace Fund, shall be established. The operations of the Peace Fund shall be governed by the relevant Financial Rules and Regulations of the Union.

2. The Peace Fund shall be made up of financial appropriations from the regular budget of Union, including arrears of contributions, voluntary contributions from Member States and from other sources within Africa, including the private sector, civil society and individuals, as well as through appropriate fund raising activities.

3. The Chairperson of the Commission shall raise and accept voluntary contributions from sources outside Africa, in conformity with the objectives and principles of the Union.

4. There shall also be established, within the Peace Fund, a revolving Trust Fund. The appropriate amount of the revolving Trust Fund shall be determined by the relevant Policy Organs of the Union upon recommendation by the Peace and Security Council.

Assessment of Cost of Operations and Pre-financing

5. When required, and following a decision by the relevant Policy Organs of the Union, the cost of the operations envisaged under Article 13(3) of the present Protocol shall be assessed to Member States based on the scale of their contributions to the regular budget of the Union.

6. The States contributing contingents may be invited to bear the cost of their participation during the first three (3) months.

7. The Union shall refund the expenses incurred by the concerned contributing States within a maximum period of six (6) months and then proceed to finance the operations.

Article 22: Final Provisions

Status of the Protocol in Relation to the Cairo Declaration

1. The present Protocol shall replace the Cairo Declaration.

2. The provisions of this Protocol shall supercede the resolutions and decisions of the OAU relating to the Mechanism for Conflict Prevention, Management and Resolution in Africa, which are in conflict with the present Protocol.

Signature, Ratification and Accession

3. The present Protocol shall be open for signature, ratification or accession by the Member States of the Union in accordance with their respective constitutional procedures.

4. The instruments of ratification shall be deposited with the Chairperson Commission.

Entry into Force

5. The present Protocol shall enter into force upon the deposit of the instruments of ratification by a simple majority of the Member States of the Union.

Amendments

6. Any amendment or revision of the present Protocol shall be in accordance with the provisions of Article 32 of the Constitutive Act.

Depository Authority

7. This Protocol and all instruments of ratification shall be deposited with the Chairperson of the Commission, who shall transmit certified true copies to all

Member States and notify them of the dates of deposit of the instruments of ratification by the Member States and shall register it with the United Nations and any other Organization as may be decided by the Union.

Adopted by the 1st Ordinary Session of the Assembly of the African Union

Durban, 9 July 2002

1. People's Democratic Republic of Algeria

2. Republic of Angola

3. Republic of Benin

4. Republic of Botswana

5. Burkina Faso

6. Republic of Burundi

7. Republic of Cameroon

8. Republic of Cape Verde

9. Central African Republic

10. Republic of Chad

11. Islamic Federal Republic of the Comoros

12. Republic of the Congo

13. Republic of Côte d'Ivoire

14. Democratic Republic of Congo

15. Republic of Djibouti

16. Arab Republic of Egypt

17. State of Eritrea

18. Federal Democratic Republic of Ethiopia

19. Republic of Equatorial Guinea

20. Republic of Gabon

21. Republic of The Gambia

22. Republic of Ghana

23. Republic of Guinea

24. Republic of Guinea Bissau

25. Republic of Kenya

26. Kingdom of Lesotho

27. Republic of Liberia

28. Great Socialist People's Libyan Arab Jamahiriya

29. Republic of Madagascar

30. Republic of Malawi

31. Republic of Mali

32. Islamic Republic of Mauritania

33. Republic of Mauritius

34. Republic of Mozambique

Appendix 225

35. Republic of Namibia

36. Republic of Niger

37. Federal Republic of Nigeria

38. Republic of Rwanda

39. Sahrawi Arab Democratic Republic

40. Republic of Sao Tome and Principe

41. Republic of Senegal

42. Republic of Seychelles

43. Republic of Sierra Leone

44. Republic of Somalia

45. Republic of South Africa

46. Republic of Sudan

47. Kingdom of Swaziland

48. United Republic of Tanzania

49. Republic of Togo

50. Republic of Tunisia

51. Republic of Uganda

52. Republic of Zambia

53. Republic of Zimbabwe

Bibliography

Acker, J. C, and Mbiti, I. M. (2010). *Mobile Phones and Economic Development in Africa.* Washington, DC: Center for Global Development.

Ackerman, Susan Rose (1999). *Corruption and Government: Causes, Consequences and Reform.* Cambridge: Cambridge University Press.

Africa Regional Integration Index, Report 2016, African Union, African Development Bank and United Nations Economic Commission for Africa, 2016.

African Development Bank Group, http://www.afdb.org/en/about-us/mission-strategy/financing-the-strategy/

Agenor, P. R. (2010). "A Theory of Infrastructure-led Development," *Journal of Economic Dynamics and Control 34*(5).

Akobeng, Eric (2016). "Gross Capital Formation, Institutions and Poverty in Sub-Saharan Africa," *Journal of Economic Policy Reform*, January 25.

Akobeng, Eric (2016. "Gross Capital Formation, Institutions and Poverty in Sub-Saharan Africa." *Journal of Economic Policy Reform*, January, 25. http://www.tandfonline.com/doi/full

Al-Khatib, Jamal A., Avinash Malshe, and John J. Sailors (2011). "The Impact of Deceitful Tendencies, Relativism and Opportunism on Negotiation Tactics: A Comparative Study of US and Belgian Managers." *European Journal of Marketing*, 45(½).

Allott, A. (1980). *The Limits of Law.* London: Butterworths.

Almond, M. A., and S. Syfert (1997). "Beyond Compliance: Corruption, Corporate Responsibility and Ethical Standards in the New Global Economy." *North Carolina Journal of International Law and Commercial Regulation 22.*

Alston, Philip, and Mary Robinson (Eds.) (2005). *Human Rights and Development: Towards Mutual Reinforcement.* Oxford: Oxford University Press.

Amartya K. Sen (1983). "Development Which Way Now?" *Economic Journal (December).*

Amin A. (1999). An institutionalist perspective on regional economic development. *International Journal of Urban and Regional Research 23*.

Ammer, Christine, and Dean S. Ammer (1977). *Dictionary of Business and Economics*. New York: The Free Press.

Amutabi, Maurice N. (2006). *The NGO Factor in Africa: The Case of Arrested Development in Kenya*. New York: Routledge.

Anand, B. H., and T. Khanna (2000). "Do Firms Learn to Create Value? The Case of Alliances." *Strategic Management Journal, 21*(3).

Anderson, E., and H. Gatignon (1986). Modes of foreign entry. A transaction cost analysis and propositions. *Journal of International Business Studies, 17*(3).

Antonio Maria Costa (2007). *Anti-Corruption Climate Change: It Started in Nigeria*, United Nations Office on Drugs and Crime, 6th National Seminar on Economic Crime. Abuja, Nigeria, Nov. 13. http://www.unodc.org/unodc/en/frontpage/nigerias-corruption-busters.html

Anyanwu, J. C. (2012). "Why Does Foreign Direct Investment Go Where It Goes?: New evidence from African countries." *Annals of Economics and Finance 13*(2).

Arcudi, Giovanni, and M. E. Smith (2013). "The European Gendarmerie Force: A Solution in Search of Problems?" *European Security, 22*(1).

Arikan, Ilgaz, and O. Shenkar (2013). "National Animosity and Cross-Border Alliances," *Academy of Management Journal, 56*(6) (December).

Armstrong H. W., and J. Taylor (2000). *Regional Economics and Policy*. Blackwell, Oxford.

Arnold C. Herberger (1983). "The Cost-Benefit Approach to Development Economics." *World Development, 11*(10).

Arpan, J. S. (1972). *International Intracorporate Pricing: Non-American Systems and Views*, New York: Praeger.

Arrow, Kenneth (1962). "The Economic Implications of Learning by Doing." *Review of Economic Studies* (June).

Arthur, Lewis W. (1984). "The State of Development Theory." *American Economic Review* (March).

Ascani, Andrea, Riccardo Crescenzi, and Simona Iammarino (2012). "Regional Economic Development: A Review" *Search Working Paper*, WP1/03, European Commission, January.

AU/NEPAD African Action Plan 2010–2015: Advancing Regional and Continental Integration in Africa, African Union, October 17, 2009.

Ayittey, George B. N. (1989). "The Political Economy of Reform in Africa." *Journal of Economic Growth*, (Spring).

Azzoni C. (2001). Economic growth and regional income equality in Brazil. *The Annals of Regional Science 35.*

Balaam, David N., and Michael Veseth, (Eds.) (1996). *Readings in International Political Economy,* Upper Saddle River, NJ: Prentice-Hall.

Banks, Ken (2015). "The Top-Down, Bottom-Up Development Challenge." *Stanford Social Innovation Review,* (December).

Barclay, L. A., and B. Oyelaran-Oyeyinka, (2004). "Human Capital and Systems of Innovation in African Development," *African Development Review 16*(1).

Barkat Aftab Ahmed Khalid Ahmed (2014). "Role of Corruption in Human Rights Violation." *Scholarly Research Journal for Humanity Science & English Language, I*(IV), (June–July).

Bassiouni, M. C. (1999). *International Crime Law: Crimes.* Vol. 1, (2nd ed.). New York: Transnational Publishers.

Batley, R. and R. Shah (2009). "Private-Sector Investment in infrastructure: Rationale and Causality for Pro-poor Impacts." *Development Policy Review 27*(4).

Beavis, Mark (1999–2014). "The IR Theory Knowledge Base: IR Paradigms, Approaches and Theories," irtheory@hotmail.com.

Beddies, C. (1999). *"Investment Capital Accumulation and Growth-Some Evidence from The Gambia 1964–98."* Washington, DC: International Monetary Fund.

Beets, S. D. (2005). "Understanding the Demand-Side Issues of International Corruption." *Journal of Business Ethics 57*(1).

Ben-David, D. (1998), "Convergence Clubs and Subsistence Economies." *Journal of Development Economics,* 55.

Beresford, M. (2008). "Doi Moi in Review: The Challenges of Building Market Socialism in Vietnam." *Journal of Contemporary Asia, 38*(2).

Berger, P. L. (1974). *Pyramids of Sacrifice.* New York: Basic books.

Bernstein, H. (1992). "Poverty and the Poor." In *Rural Livelihoods: Crisis and Responses,* H. Bernstein, B. Crown and H. Johnson (Eds.). Oxford: Oxford University Press.

Bhagwati, J. (1984). "Development Economics: What Have We Learned?" *Asian Development Review, 2*(1).

Bhargava, Sandeep (1994). "Profit Sharing and the Financial Performance of Companies: Evidence from U.K. Panel Data." *Economic Journal, 104.*

Blake, R. C. (2000). "The World Bank's Draft Comprehensive Development Framework and the Micro-Paradigm of Law and Development." *Yale Human Rights and Development Law Journal 3.*

Bibliography

Blinder, Alan S. (Ed.) (1989). *Paying for Productivity: A Look at the Evidence*, Washington, DC: The Brookings Institute.
Bliss, C. J. (1975). *Capital Theory and the Distribution of Income*, New York: American Elsevier Publishing Company, Inc., 1975.
Bobo, Benjamin F. (1981). *Corporate and Third World Involvement: A Reciprocal Relationship*, unpublished manuscript (University of California, Los Angeles and University of California, Riverside), 421 pages.
Bobo, Benjamin F. (2001), *Locked In and Locked Out: The Impact of Urban Land Use Policy and Market Forces on African Americans*. Westport, Connecticut: Praeger Publishers.
Bobo, Benjamin F. (2005). *Rich Country, Poor Country: The Multinational as Change Agent*. Westport, Connecticut: Praeger Publishers.
Bobo, Benjamin F. (2012). "Implementing State Interventionist Development: The Role of the Multinational Corporation." In Benjamin F. Bobo, and Herman Sintim-Aboagye (Eds.), *Neo-Liberalism, Interventionism and the Developmental State: Implementing the New Partnership for Africa's Development*. Trenton New Jersey: Africa World Press.
Bobo. Benjamin F. and Hermann Sintim-Aboagye (Eds.) (2012). *Neo-Liberalism, Interventionism and the Developmental State: Implementing the New Partnership for Africa's Development*. Trenton, New Jersey: Africa World Press.
Bodammer, I., Pirie, M. F., and Addy-Nayo, C. (2006). Telecoms regulation in developing countries: Attracting investment into the sector: Ghana-A case study. *Journal of World Trade*, 39(3).
Botchway, Karl, and Jamee Moudud (2012). *Neo-Liberalism and the Developmental State: Consideration for the New Partnership for Africa's Development"* Chapter 2. In Benjamin F. Bobo, and Hermann Sintim-Aboagye (Eds.), *Neo-Liberalism, Interventionism and the Developmental State: Implementing the New Partnership for Africa's Development*, Trenton, New Jersey: Africa World Press.
Brand, Ronald A. (2000). Fundamentals of International Business Transactions. The Hague, The Netherlands: Kluwer Law International.
Brander, James A. (1990). "Rationales for Strategic Trade and Industrial Policy." In Paul R. Krugman, (Ed.). *Strategic Trade Policy and the New International Economics*, Cambridge, Massachusetts: MIT Press.
Braselle, Anne-Sophie, Gaspart, Frederic, and Platteau, Jean-Phillipe (2002). "Land Tenure Security and Investment Incentives: puzzling Evidence from Burkina Faso." *Journal of Development Economics*: 67(2).

Browne, Robert S. (1996). "How Africa Can Prosper." In David N. Balaam and Michael Veseth (Eds.), *Readings in International Political Economy*. Upper Saddle River, New Jersey: Prentice-Hall, Inc.

Budd, L., and Hirmis, A. K. (2004). Conceptual Framework for Regional Competitiveness. *Regional Studies*, *38*(9).

Buehn, Andreas, and Stefan Eichler (2009). "Trade Misinvoicing: the Dark Side of World Trade." EEA 2009) www.Eea-esem.com/files/papers/EEA-ESEM/2009/1065/Buehn-Eichler.pdf.

Byamugisha, Frank F. K. (2013). *Securing Africa's Land for Shared Prosperity: A Program to Scale Up Reforms and Investments*. Africa Development Forum series. Washington, DC: World Bank. doi: 10.1596/978-0-8213-9810-4. License: Creative Commons Attribution CC BY 3.0.

Byerlee, D., A. De Janvy, and E. Sadoulet (2009). "Agriculture for Development: Towards a New Paradigm." *Annual Review of Resource Economics 1*(1).

Caglar, K., and P. K. O'Brien (1978). *Economic Growth in Britain and France 1780–1914: Two Paths to the Twentieth Century*. London: Allen & Unwin.

Calderon, Cesar (2009). Infrastructure and Growth in Africa. *Policy Research Working Paper 4914*, The World Bank, Africa Region, African Sustainable Development Office, (April).

Calitz, E., and J. Fourie (2010). "Infrastructure in South Africa: Who Is to Finance and Who Is to Pay?" *Development Southern Africa 27*(2).

Canada-China Economic Complementarities Study (2013). *Economic Partnership Working Group* Foreign Affairs, Trade and Development Canada, Sept. www.international.gc.ca./trade-agreements.

Cantwell, J., and S. Iammarino (2003). *Multinational Corporations and European Regional Systems of Innovation*. Routledge, London.

Carbaugh, Robert J. (2002). International Economics (8th ed.).Cincinnati, Ohio: South-Western.

Carliner, Geoffrey (1990). "Industrial Policies for Emerging Industries." In Paul R. Krugman (Ed.). *Strategic Trade Policy and the New International Economics*. Cambridge, Massachusetts: MIT Press.

Carothers, Thomas (2006). "The Problem of Knowledge." In *Promoting the Rule of Law Abroad, In Search of Knowledge* Thomas Carothers (Ed.). Washington D.C: Carnegie Endowment for International Peace.

Carr Center for Human Rights. *Measurement and Human Rights: tracking Progress, Assessing Impact*. Policy Project Report, Harvard University, 2005. www.hks.harvard.edu/cchrp/pdf/Measurement 2005 Report.pdf

Cernea, M. M. (1984). Can Local Participation Help Development? Mexico's PIDER program shows it can improve the selection and execution of local development projects. *Finance & Development, 21*(4).

Chang, Ha-Joon (2001). "Institutional Development in Developing Countries in a Historical Perspective: Lessons from Developed Countries in Earlier Times." Unpublished paper, University of Cambridge.

Chang, Ha-Joon (2001). "Institutional Development in Developing Countries in a Historical Perspective: Lessons from Developed Countries in Earlier Times." Unpublished paper, University of Cambridge.

Chapra, Sunil, and Peter Meindl (2004). *Supply Chain Management* (2nd ed.). Upper Saddle River: Pearson Prentice Hall.

Chen, H., and Rozelle, S. (1999). "Leaders, Managers, and the Organization of Township and Village Enterprises in China," *Journal of Development Economics 60*(2).

Chenery, Hollis (1986). "Growth and Transformation." In Hollis Chenery, Sherman Robinson, and Moshe Syrquin (Eds.), *Industrialization and Growth*. New York: Oxford University Press.

Cheshire P., and S. Magrini (2000). Endogenous process in European regional growth: Convergence and policy. *Growth and Change 31*.

Chetwynd, Eric, Frances Chetwynd, and Bertram Spector (2003). *Corruption and Poverty: A Review of Recent Literature.* Management Systems International, www.u4.no/pdf/?file=/document/literature/corruption-and-poverty.pdf

CIA World Factbook – The Best Country factbook available online, www.ciaworldfactbook.us/.

Clarke, D. (1999). "The Meany Meanings of the Rule of Law." In *Law, Capitalism and Power in Asia*. K. Jayasuriya (Ed.). London: Routledge.

Coase, Ronald, and Ning Wang (2012). *How China Became Capitalist*. New York: Palgrave Macmillan.

Coe D., and E. Helpman (1995). International R&D spillovers. *European Economic Review*.

Collier, P., and J. W. Gunning (1999). "Explaining African Economic Performance." *Journal of Economic Literature, 37*(March).

Conover, Steve (2012). 'Top-Down' vs. 'Bottom-Up,' *The American* (September), https://www.aei.org/publication/top-down-vs-bottom-up/ Constitution of the United States, http://www.senate.gov/civics/constitution/item/constitution.htm

Constitutive Act of the African Union, (2000). http://www.au.int/en/sites/default/files/ConstitutiveActEN.pdf

Convention on the Non-Applicability of Statutory Limitations to War Crimes and Crimes Against Humanity (1970). Office of the High Commissioner for Human Rights, New York: United Nations, November 11, http://www.ohchr.org/EN/ProfessionalInterest/Pages/WarCrimes.aspx

Conway, G., and Waage, J. (2010). *Science and Innovation for Development.* London: UK Collaborative on Development Sciences.

Corbett, S. (2008). Can the cellphone help end global poverty? *New York Times Magazine* (April 13). http://www.nytimes.com/2008/04/13/magazine/13anthropology-t.html

Cornwell, Richard (2002). "A New Partnership for Africa's Development?" *African Security Review*, *11*(1).

Cost of U.S. Overseas Military Bases (2012). LessWaiting.com, (April 23). http://www.globalsecurity.org/military/facility/images/diego-garcia-ims7.jpg

Costinot, Arnaud, Jonathan Vogel, and Su Wang (2013). "An Elementary Theory of Global Supply Chains," *Review of Economic Studies*, *80*.

Cotula, Lorenzo (Ed.) (2007). *Changes in "Customary" Land Tenure Systems in Africa.* International Institute for Environment and Development, SMI (Distribution Services) Ltd: Stevenage, Hertfordshire.

Crafts N. (2004). Globalization and Economic Growth: A historical perspective. *The World Economy 27.*

Crescenzi R., A. Rodriguez-Pose (2011). Reconciling top-down and bottom-up development policies. *Environment and Planning A, 43.*

Crescenzi, Riccardo, and Andrés Rodríguez-Pose (2011). "Reconciling Top-Down and Bottom-Up Development Policies," *Working Paper Series in Economic and Social Sciences 2011/03*, IMDEA Social Sciences Institute, (January).

Crooks, E. (2009). "Shell to Lend Nigeria $3bn," *Financial Times, FT.com*, (February).

Curry, Jr., Robert L. and Donald Rothchild, (1974). "On Economic Bargaining between African Governments and Multinational Companies," *Journal of Modern African Studies*, *12*(2).

David, G. (1993). Strategies for grass roots human development. *Social Development Issues, 15*(2).

Davis, Kevin E., and Michael J. Trebilcock (2001). "Legal Reforms and Development." *Third World Quarterly 22*(1).

Davoodi, Hamid, R. Sanjeev Gupta, and Erwin R. Tiongson (2000). *Corruption and the Provision of Health Care and Education Services.* IMF Working Paper No. 00/116, 2000. www.imf.org/external/pubs/ft/wp/2000/wp00116.pdf.

De Ferranti, D., and A. J. Ody (2007). Beyond microfinance: Getting capital to small and medium enterprises to fuel faster development. *The Brookings Institution Policy Brief #159*. http://www.brookings.edu/articles/2007/03 developmentdeferranti.aspx

De Soto, Hernando (2000). *The Mystery of Capital, Why Capitalism Triumphs in the West and Fails Everywhere Else*. London: Black Swan.

Dezelay, Y., and Garth, B. (2001). "The Import and Export of Law and Legal Institutions: International Strategies in National Palace Wars." In D. Nelken and J. Feest (Eds.), *Adapting Legal Cultures*. Oxford: Hart Publishing.

Di Tella, R., and Savedoff, W. D. (Eds.) (2001). *Diagnosis Corruption: Fraud in Latin America's Public Hospitals*. Inter-American Development Bank.

Diamond, Adam, James Barham, and Debra Tropp (2009). "Emerging Market Opportunities for Small-Scale Producers," Proceedings of a Special Session at the USDA Partners Meeting, U.S. Department of Agriculture, Agricultural Marketing Service (April) http://x.doi.org/10.9752/MS034.04-2009.

Dick, H. (2007). "Why Law Reform Fails, Indonesia's Anti-Corruption Efforts." In T. Lindsey (Ed.), *Law Reform in Developing and Transitional States*. New York: Routledge.

Dollar, David (2015). United States-China Two-Way Direct Investment: Opportunities and Challenges. Brookings Institution, (January).

Dollar, David, and Kraay Aart. (2004). "Trade, Growth, and Poverty." *The Economic Journal*, *114*(439).

Donner, J. (2008). Research Approaches to Mobile Use in the Developing World: A review of the literature. *The Information Society*, *24*.

Dwarkasing, Ramon (2011). Associated Enterprises: A Concept Essential for the Application of the Arm's Length Principle and Transfer Pricing. The Netherlands: Dwarkasing & Partners (Management@dwarkasing.com).

Dwarkasing, Ramon (2013). "Comments from Academia on the Revised Discussion Draft on Transfer Aspects of Intangibles." The Netherlands: Maastricht University, September 27.

Easterly, William (2008). "Institutions: Top Down or Bottom Up?" *American Economic Review*: Papers and Proceedings, *98*(2). http://www.aeaweb.org/articles.php?doi=10.1257/aer.98.2.95

Eberhard, A., and K. N. Gratwick, (2008). "An Analysis of Independent Power Projects in Africa: understanding Development and Investment Outcomes," *Development Policy Review* *26*(3).

Edwards, S. (2002). Information technology and economic growth in developing countries. *Challenge, 45*(3).
Emerson K., T. Nabatchi, and S. Balogh (2012). "An integrative framework for collaborative governance." *Journal of Public Administration Research and Theory 22.*
Enowbi Batuo, Michael (2008). "The Role of Telecommunication Infrastructure in the Regional Economic Growth of Africa." http://mpra.ub-muenchen.de/12431/, MPRA Paper No. 12431, posted 30. December 2008/18:47, 22. June.
Escribano, Alvaro, J. Luis Guasch. and Jorge Pena (2008). "Impact of Infrastructure Constraints on Firm Productivity in Africa," Working Paper 9, Africa Infrastructure Sector Diagnostic, World Bank, Washington, D.C.
Esfahani, H. S., and M. T. Ramírez (2003). "Institutions, Infrastructure, and Economic Growth." *Journal of Development Economics 70*(2).
Extracting Justice: Battling Corruption in Resource-Rich Africa, Devex, August 14, 2014, https://www.devex.com/news/extracting-justice-battling-corruption-in-resource-rich-africa-84137
Eyraud, L. (2009). "Why isn't South Africa Growing Faster? A Comparative Approach." *International Monetary Fund.*
Fagerberg, Jan (1988). "Why Growth Rates Differ." In Giovanni Dosi et al. (Eds.), *Technical Change and Economic Theory*. New York: Pinter.
Farrell, G., and S. Isaacs (2007). Survey of ICT and education in Africa: A summary report, based on 53 country surveys. Washington, D.C: infoDev/World Bank.
Farrell, M. (2009). "EU Policy Towards Other Regions: Policy learning in the External Promotion of Regional Integration." *Journal of European Public Policy 16*(8).
Faundez, Julio (2001). "Legal Reform in Developing and Transitional Countries." In Rudolph V. Van Puymbroeck (Ed.), *Comprehensive Legal and Judicial Development*. Washington: World Bank.
Felstiner, W., R. Abel, and A. Sarat, (1980). "The Emergence and Transformation of Disputes: Naming, Blaming, Claiming." *Law and Society Review 15.*
Ferreria, Alejandrino J. (2001). "Do Good While Doing Well." http://www.philstar.com/business-usual/137504/do-good-while-doing-well, October.
FitzRoy Felix, and Kornelius Kraft (1986). "Profitability and Profit Sharing." *Journal of Industrial Economics, 35.*
Flint, Yan (2011). "Capitalism Vietnamese-style: Combining Top-down with Bottom-up." *Rethinking Development in an Age of Scarcity*

and Uncertainty. EADI/DSA, University of York, September 19–22, http://eadi.org/gc2011/flint-625.pdf

Flint, Yan, (2008), Human Development Reports, United Nations Development Program, 2015, http://hdr.undp.org/en/composite/HDI, and Human Development Index 1975–2005 – Country Rankings, 2008.

Foreign direct investment flows. *OECD Factbook 2015–2016. Economic, Environmental and Social Statistics*, Paris: OECD Publishing.

Foster, Vivien, and Cecilia Briceno-Garmendia, (Eds.) (2010). *Africa's Infrastructure: A Time for Transformation,* Washington, DC: The International Bank for Reconstruction and Development/The World Bank.

Frank, Gunder Andre (1996). "The Development of Underdevelopment." In David N. Balaam and Michael Veseth (Eds.), *Readings in International Political Economy*, Upper Saddle River, NJ: Prentice-Hall.

Freeman, C., and F. Louca (2001). *As Time Goes By: From the Industrial Revolution to the Information Revolution.* Oxford: Oxford University Press.

French National Gendarmerie, http://www.fiep.org/member-forces/french-national-gendarmerie *and* http://www.gendarmerie.interieur.gouv.fr

Fulcher (2004). *Capitalism* (1st ed.). New York: Oxford University Press.

Gabel, Medard, and Henry Bruner (2003). *Global Inc. An Atlas of The Multinational Corporation.* New York: The New Press.

Garbacz, C., and H. G. Thompson, (2007). Demand for telecommunication services in developing countries. *Telecommunication Policy, 31*(5).

Garza, R. T., S. A. Isonio, and P. I. Gallegos (1988). Community development in rural Mexico: The social psychological effects of adult education. *Journal of Applied Social Psychology 18*(8).

Gavin, B. (2005). "The Euro Mediterranean Partnership: An Experiment in North-South-South Integration." *Intereconomics 40*(6).

Gerschenkron, Alexander (1962). *Economic Backwardness in Historical Perspective.* Cambridge, Massachusetts: Harvard University Press.

Gertler, M. S., and Vinodrai, T. (2009). "Life Sciences and Regional Innovation: One Path or Many?" *European Planning Studies 17*(2).

Ghazanchyan, M., and J. G. Stotsky (2013). "Drivers of Growth: Evidence from Sub-Saharan African Countries." IMF Working Paper, WP/13/236, *International Monetary Fund,* (November).

Ghura, D. (1997). "Private Investment and Endogenous Growth-Evidence From Cameroon." IMF Working Paper, WP/97/165, *International Monetary Fund.* (December).

Ghura, D., and Michael T. Hadji (1996). Growth in Sub-Saharan Africa. *Staff Papers*, International Monetary Fund, 43, September.
Goetz, Anne Marie, and Rob Jenkins (2002). *Voice, Accountability and Human Development: The Emergence of a New Agenda.* UNDP, 2002. http://hdr.undp.org/en/reports/global/hdr2002/papers/Goetz-Jenkins_2002.pdf
Golub, Stephen (2006). "A House without Foundation." In Thomas Carothers (Ed.), *Promoting the Rule of Law Abroad, In Search of Knowledge*, Washington D.C: Carnegie Endowment for International Peace.
Goodpaster, Gary (2007). "Law Reform in Developing Countries." In T. Lindsey (Ed.), *Law Reform in Developing and Transitional States.* New York: Routledge.
Goody, Jack (1963). "Feudalism in Africa?" *Journal of African History, IV*(1).
Gopakumar, K. (1998). "Citizen Feedback Surveys to Highlight Corruption in Public Services: The Experience of Public Affairs Centre." Public Affairs Centre, Banglalore, India."
Greenberg, Dolores (1982). "Reassessing the Power Patterns of the Industrial Revolution: An Anglo-American Comparison." *American Historical Review, 87*(December).
Greene, J., and M. Duerr (1970). *Intercompany Transactions in the Multinational Firm.* New York: The Conference Board.
Griffin, Keith, and John Gurley (1985). "Radical Analyses of Imperialism, the Third World, and the Transition to Socialism: A Survey Article." *Journal of Economic Literature, 23*(September).
Grindle, Merilee S. (2002). "Good Enough Governance: Poverty Reduction and Reform in Developing Countries." Kennedy School of Government, Harvard University (November).
Grossman G., and E. Helpman (1991a). Endogenous innovation in the theory of growth. *Journal of Economic Perspectives 8.*
Grossman G., and E.Helpman (1991b). *Innovation and growth in the global economy.* Cambridge, MA: MIT Press.
Haacke, J., and P. D.Williams (2008). "Security Culture, Transnational Challenges and the Economic Community of West African Slaves." *Journal of Contemporary African Studies 26*(2).
Hanson, Stephanie (2009). *Corruption in Sub-Saharan Africa.* Council on Foreign Relations, Washington, DC. (August).
Harrigan, K. R. (1985). Vertical integration and corporate strategy. *Academy of Management Journal. 28*(2).

Harrigan, Kathryn Rudie (1985)."Strategic Alliances as Agents of Competitive Change." https://www0.gsb.columbia.edu/mygsb/faculty/research/pubfiles/10497/Harrigan%20Strategic%20Alliances.pdf

Harrigan, Kathryn Rudie (1988). "Joint Ventures and Competitive Strategies," *Strategic Management Journal*, 9(2) (March–April).

Hart, S. L., and C. K. Prahalad (2004). *The Fortune at the Bottom of the Pyramid: Eradicating Poverty through Profits*. Upper Saddle River, NJ: Wharton.

Heidenheimer, A. J., et al. (Eds.) (2002) *Political Corruption: Concepts and Texts*, New Brunswick: Transaction Publishers.

Herberger, A. C. (1983). "The Cost-Benefit Approach to Development Economics." *World Development 11*(10).

Hernández-Catá, E. (2000). "Raising Growth and Investment in Sub-Saharan Africa: What Can be Done?" *Policy Discussion Paper: PDP/00/4*, International Monetary Fund, Mayhington.

Hines, Andrew (2005). "What Human Rights Indicators Should Measure." In *Measurement and Human Rights: Tracking Progress, Assessing Impact*. Carr Center for Human Rights. Policy Project Report, Harvard University, www.hks.harvard.edu/cchrp/pdf/Measurement 2005 Report.pdf

Hirschman, Albert O. (1958). *The Strategy of Economic Development*. New Haven: Yale University Press.

Hirschman, Albert O. (1968). "Political Economy of Import Substituting Industrialization." *Quarterly Journal of Economics,* (February).

Hosman, L., and E. Fife (2008). Improving the prospects for sustainable ICT projects in the developing world. *International Journal of Media and Cultural Politics*, 4(1).

Hosman, Laura, and Elizabeth Fife (2012). "The Potential and Limits of Mobile Phone Usage for Development in Africa: Top-Down-Meets-Bottom-Up Partnering." *Journal of Community Informatics* 8(3).

Hughes, Jonathan and Jeff Weiss (2007). "Simple Rules for Making Alliances Work." *Harvard Business Review* (November).

Human Development Report (2015). United Nations Development Programme, http://hdr.undp.org

Hylton, Maria O'Brien (1992). "Socially Responsible" Investing: Doing Good versus Doing Well in an Inefficient Market, *The American University Law Review*, 42(1).

Ibrahim, Mo (2015). *Ibrahim Index of African Governance*, http://static.moibrahimfoundation.org/u/2015

Ignatieff, M., and K. Desormeau (2005). "Measurement and Human Rights Introduction." In *Measurement and Human Rights: Tracking Progress, Assessing Impact*. Carr Center for Human Right. Policy Project Report, Harvard University, 2005. www.hks.harvard.edu/cchrp/pdf/Measurement 2005Report.pdf

Ilorah, Richard (2012). "The Realities of Regional Integration: The NEPAD Perspective." In Benjamin F. Bobo, and Sintim-Aboagye (Eds.), *Neo-Liberalism. Interventionism and the Developmental State: Implementing the New Partnership for Africa's Development*. Trenton, New Jersey: Africa World Press.

Impact of Corruption on Nigeria's Economy. (2016). Pricewaterhouse Coopers, (PwC), http://www.pwc.com/ng/en/publications/impact-of-corruption-on-nigerias-economy.html

Insuring FDI's Success (2014)." *Zurich*, October 3, https://www.zurich.com/en/knowledge/articles/2014/10/insuring-fdi-success

Itagaki, Y. (1963). "Criticism of Rostow's Stage Approach: The Concepts of Stage, System and Type." *The Developing Economies, 1*.

Jayasuriya, K. (1999). "The Rule of Law and Governance in the Asian State." *Australian Journal of Asian Law 1*(2).

Jensen, M. (1998). Internet opens new markets for Africa. *Africa Recovery Online A United Nations Publication*, *12*(3). http://www.un.org/ecosocdev/geninfo/afrec/vol12no3/internet1.htm

Johnston, Michael (2005). *Syndromes of Corruption: Wealth, Power, and Democracy*. Cambridge: Cambridge University Press.

Joireman, S. F. (2008). "The Mystery of Capital Formation in Sub-Saharan Africa: Women, Property Rights and Customary Law." *World Development 36*(7).

JSTOR (2008). *Africa Today*, (review of *The NGO Factor in Africa: The Case of Arrested Development in Kenya*), *54*(4), (Summer).

Juma, C., Y. C. Lee, and T. Ridley (2006). "Infrastructure, Innovation and Development." *International Journal of Technology and Globalization 2*(3).

Juma, Celestous (2011). *The New Harvest: Agricultural Innovation in Africa*. Oxford: Oxford University Press.

Kapiga, Kipaya "Bottom-up and Top-down Approaches to Development." *Global Social Entrepreneurship*, http://global_se.scotblogs.wooster.edu/2011/06/26/bottom-up-and-top-down-approaches-to-development/

Kaplinsky, R. (2009). "Below the Radar: What Does Innovation in Emerging Economies Have to Offer Other Low-income Countries?" *International Journal of Technology Management and Sustainable Development 8*(3).

Kar, Dev, and Brian LeBlanc (2013). "Illicit Financial Flows from Developing Countries: 2002–2011." *Global Financial Integrity*. (December)

Kar, Dev, and Sarah Freitas (2012). "Illicit Financial Flows from China and the Role of Trade Misinvoicing." *Global Financial Integrity* (October).

Kennedy, David (2003). "Laws and Developments." In John Hatchard and Amanda Perry-Kessaris (Eds.), *Law and Development Facing Complexity in the 21st Century*. London: Cavendish Publishing Limited.

Kennedy, David (2006). "The Rule of Law." Political Choices and Development Common Sense." In David M. Trubeck and Alvaro Santos (Eds.), *The New Law and Economic Development, A Critical Appraisal*. Cambridge: Cambridge University Press.

Ketchen, Jr., D. J., and G. T. M. Hult (2011), "Building Theory About Supply Chain Management: Some Tools From The Organizational Sciences," *Journal of Supply Chain Management, 47*(2) (April).

Khan, M. S., and C. M. Reinhart. (1990). "Private Investment and Economic Growth in Developing Countries." *World Development 18*(1).

Kihika, Maureen (2009). "Development or Underdevelopment: The Case of Non-Governmental Organizations in Neoliberal Sub-Saharan Africa." *Journal of Alternative Perspectives in the Social Sciences*, *1*(3) (December).

Kim, Seongsu (1998). "Does Profit Sharing Increase Firm's Profits?" *Journal of Labor Research*, *19*(2).

Kipaya, Kapiga (2011). *"Bottom-up and Top-down Approaches to Development,"* Global Social Entrepreneurship, http://globalse.scotblogs.wooster.edu/2011/06/26/bottom-up-and-top-down-approaches-to-development/

Kitgaard, Robert (1988). *Controlling Corruption*. Berkley: University of California Press.

Klein, Peter (2014). "Bottom-up Approaches to Economic Development," *Organizations and Markets,* (November). https://organizationsandmarkets.com/2014/11/19/bottom-up-approaches-to-economic-development/

Kleinfeld, Rachel (2006). "Competing Definitions of the Rule of Law." In Promoting the Rule of Law Abroad. In Thomas Carothers (Ed.), *Search of Knowledge*. Washington D.C: Carnegie Endowment for International Peace.

Kohler, Daniel F. (1988). *The Effects of Defense and Security on Capital Formation in Africa*, Santa Monica: The Rand Corporation, N-2653-USDP, (September); and https://www.rand.org/content/dam/rand/pubs/notes/2007/N2653.pdf

Koontz T. M., and J. Newig (2014). From planning to implementation: Top-down and bottom-up approaches for collaborative watershed management. *Policy Studies Journal 42*(3).

Krugman P. and Elizondo R. Livas (1996). "Trade policy and the Third World Metropolis." *Journal of Development Economics 49*.

Krugman, Paul R. (1986). *Strategic Trade Policy and the New International Economics*. Massachusetts: MIT Press.

Kuznets, Simon (1952). "Proportion of Capital Formation to National Product "*The American Economic Review*, *42*(2). Papers and Proceedings of the Sixty-fourth Annual Meeting of the American Economic Association (May) http://data.worldbank.org; and http://lexicon.ft.com

Kyem, P. A. K., and P. K. LeMaire (2006). Transforming recent gains in the digital divide into digital opportunities: Africa and the boom in mobile phone subscriptions. *The Electronic Journal on Information Systems in Developing Countries*, *28*(5).

Landes, David S. (1990). "Richard T. Ely Lecture." *American Economic Review Papers and Proceedings 80*(2) (May).

Langmia, Kehbuma (2011). "The Secret Weapon of Globalization: China's Activities in Sub-Saharan Africa." *Journal of Third World Studies*, *XXVIII*(2).

Le Pere, Garth, and Francis Ikome (2009). "Challenges and Prospects for Economic Development in Africa." *Asia-Pacific Review, 16*(2) (November).

Leblanc, Brian (2014). "Africa: Trade Misinvoicing, or How to Steal From Africa." allAfrica.com: Africa: Trade Misinvoicing, or How to Steal From Africa, (May).

Legrain, Philippe (2002). "Africa's Challenge," *World Link, 15*(3), (May/June).

Leibenstein, Harvey (1957). *Economic Backwardness and Economic Growth*. New York: John Wiley & Sons.

Lesay, Ivan (2012). "How 'Post' is the Post-Washington Consensus?" *Journal of Third World Studies*, *XXIX* (2).

Lewis, D., and T. Wallace (2000). *New Roles and Relevance: Development NGOs and the Challenge of Change*. Boulder, CO: Kumarian Press.

Lewis, W. Arthur (1984). "The State of Development Theory." *American Economic Review*, *2*(March).

Libecap, G. D. (1993). *Contracting for Property Rights*. Cambridge: Cambridge University Press.

Lindsey, Tim (2007), "Preface." In *Legal Reform in Developing and Transitional States*, edited by Tim Lindsey. New York: Routledge.

London, T., and S. L. Hart (2004). "Reinventing strategies for emerging markets beyond the transnational model." *Journal of International Business Studies*, 35.

Lostumbo, Michael J. et al. (2013). *Overseas Basing of U.S. Military Forces: An Assessment of Relative Costs and Strategic Benefits*. Santa Monica, CA: RAND Corporation.

Lu, H., Miethe, D. T., and B. Liang (2009). *China's Drug Practices and Policies*. Oklahoma City: Rutledge.

Lucas R. E. (1988). "On the mechanics of economic development." *Journal of Monetary Economics* 22.

Luo, Y. (2007). "Are Joint Venture Partners More Opportunistic in a More Volatile Environment? *Strategic Management Journal*, 28(1).

Macdonald, L. (1995). "NGOs and the problematic discourse of participation: Cases from Costa Rica." In D. B. Moore, and G. J. Schmitz's (Eds.), *Debating development discourse: Institutional and popular perspectives*. New York: St. Martin Press Inc.

Madamombe, I. (2005). "Energy Key to Africa's Prosperity: Challenges in West Africa's quest for electricity." *Africa Renewal*, 18:4. http://www.geni.org/globalenergy/library/media_coverage/africa-renewal/energy-key-to-africas-prosperity.shtml

Magaretta, J. (2002). "Why business models matter." *Harvard Business Review*, 80(5) (May 7).

Manning, D. (1999). *The Role of Legal Services Organizations in Attacking Poverty*. Washington D.C: World Bank.

Manuel, Trevor A. (2003). "Africa and the Washington Consensus: Finding the Right Path." *Finance & Development (The International Monetary Fund)*, 40(3) (September).

Marais, Marina (2002). *Technology: Linking Africa to the Information Superhighway*, http://www.ipsnews.net/2002/05/technology-linking-africa-to-the-information-superhighway/

Markusen, James R. (1995). "The Boundaries of Multinational Enterprise and the Theory of International Trade." *Journal of Economic Perspectives*, 9(2) (Spring).

Mas, I. (2009). "The Economics of Branchless Banking." *Journal of Monetary Economics* 4(2).

Maskell, P. (2001). "Towards a Knowledge-Based Theory of the Geographical Cluster." *Industrial and Corporative Change 10*(4).

Mattli, Walter (1999). "Explaining Regional Integration Outcomes." *Journal of European Public Policy*, 6(1), (March).

Mauro, Paolo (1995). "Corruption and Growth," *Quarterly Journal of Economics. 110*(3), (August).

Mbeki, Thabo (2003). "Mbeki: African Union is the Mother, NEPAD is Her Baby." *New African.*

McCann P. and Z. Acs (2009). Globalization: Countries, cities and multinationals. *Jena Economics Research Papers* No. 042.

McCormick, P. K. (2002). Internet access in Africa: A critical review of public policy issues. *Comparative Studies of South Asia, Africa, and the Middle East*, 22(1).

Meier, Gerald M. (1995). *Leading Issues in Economic Development*, (6th ed). NY: Oxford University Press.

Meier, Gerald M. (1995). *Leading Issues in Economic Development*. New York: Oxford University Press.

Migot-Adholla, Shem E., and Frank Place. (1998). "The Economic Effects of Land Registration on Smallholder Farms in Kenya: Evidence from Nyeri and Kakamega District." *Land Economics*, 74(3).

Mills, Greg, and Jonathan Oppenheimer (2002). "Partners Not Beggars." *Time Europe*, 160(2).

Milton Friedman (1962). *Capitalism and Freedom* Chicago: University of Chicago Press.

Mindzie, M. A. (2015). "Citizen Participation and the Promotion of Democratic Governance in Africa." *Great Insights Magazine*, 4(3) (April/May). http://ecdpm.org/great-insights/rising-voices-africa/citizen-participation-promotion-democratic-governance-africa/

Mitchell, Daniel J. B., David Lewin, and Edward E. Lawler, III, (1989). "Alternative Pay Systems, Firm Performance and Productivity." In Alan S. Blinder (Ed.), *Paying for Productivity: A Look at the Evidence.* Washington, D.C: The Brookings Institute.

Mo, Pak Hung (2001). "Corruption and Economic Growth." *Journal of Comparative Economics*, 29, (March).

Mok, K. H. (2008). "When Socialism Meets Market Capitalism: challenges for privatizing and marketizing education in China and Vietnam." *Policy Futures in Education, 6*(5).

Montesh Moses and Vinesh Basdeo (2012). "The Role of the South African National Defense Force in Policing," *South African Journal of Military Studies*, 40(1).

Moody-Stuart, G. (1997). Grand Corruption: *How Business Bribes Damage Developing Countries.* Oxford: WorldView Publishing.

Morgan K. (1997). The learning region: Institutions, innovation and regional renewal. *Regional Studies 31.*

Morris, E. J. (2008). "The Cartagena Protocol: Implications for Regional Trade and Technology Development in Africa," *Development Policy Review, 26*(1).

Movery, D. C., J. E. Oxley, and B. S. Silverman (1996). Strategic alliances and interfirm knowledge transfer. *Strategic Management Journal, 17.*

Movery, D. C., J. E. Oxley, and B. S. Silverman (2000). Strategic Alliances and Interfirm Knowledge Transfer." *Strategic Management Journal, 17.*

Muchie, Mammo (2002). "Wanted: African Single Currency." *New African, 407*(May).

Muckalia, Domingos Jardo (2004). "Africa and China's Strategic Partnership," *Africa Security Review. 13*(1).

Mushtaq, H. K. (1999). *Governance and Anti-Corruption Reforms in Developing Countries: Policies, Evidence and Ways Forward.* www.policyinnovations.org/ideas/policylibrary/data/01348/res/id+saFile1/1352KhanCorruption.pdf

Myrdal, Gunnar (1957). *Economic Theory and Underdeveloped Regions.* Bombay, India: Vora & Co. Publishers Private, LTD.

Myrdal, Gunnar (1968). Asian Drama: An Inquiry into the Poverty of Nations. New York: Penguin Press.

Nafziger E. Wayne (1990). *The Economics of Developing Countries*, (2nd ed.). Englewood Cliffs, NJ: Prentice-Hall.

Nafziger, E. Wayne (1990). *The Economics of Developing Countries*, (2nd ed.). Englewood Cliffs, NJ: Prentice-Hall.

Ndikumana, L. (2000). "Financial Determinants of Domestic Investment in Sub-Saharan Africa." *World Development, 28*(2).

Nee, Victor (2010). "Bottom-up Economic Development and the Role of the State." *CSES Working Paper Series*, Paper #49, Cornell University: Center for the Study of Economy & Society, (January), http://people.soc.cornell.edu/nee/pubs/bottomupecondevandstate.pdf

Neimark, Benjamin D. (2013). "The Land of Our Ancestors: Property Rights, Social Resistance, and Alternatives to Land Grabbing in Madagascar." LDPI Working Paper 26, The Land Deal Politics Initiative, www.iss.nl/ldpi landpolitics@gmail.com, (March).

Nel, E. (2001). Local Economic Development: A Review and Assessment of its Current Status in South Africa. *Urban Studies 38*(7).

Nelson, R. (2008). "What Enables Rapid Economic Progress: What Are the Needed Institutions?" *Research Policy 37*(1).

Nelson, R. R. (1956). "A Theory of the Low-Level Equilibrium Trap in Underdeveloped Economies." *American Economic Review,* (December).

Nevin, Edward (1961). *Capital Funds in Underdeveloped Countries*, New York: St. Martin's Press.

New Partnership for Africa's Development (NEPAD), October 2001.

Nhamo, Godwell, and Caiphas Chekwoti (Eds.) (2014). *Land Grabs in a Green African Economy: Implications for Trade, Investment and Development Policy*, Pretoria, South Africa: Africa Institute of South Africa.

Nigeria's Corruption Busters (2007). United Nations Office on Drugs and Crime, November 20.

Nizamuddin, Ali M. (2007), "Multinational Corporations and Economic Development: The Lessons of Singapore," *International Social Science Review*, *82*(3/4).

Njoroge, Dennis Wanyenji (2004).*Grand Corruption as a Crime Against Humanity*, http://kenyalaw.org/kl/fileadmin/pdfdownloads/Moi_B_RE SEARCH.pdfames

Nourse, Edwin G. (1944). *Price Making in a Democracy*. Washington, DC: The Brookings Institution.

Nurkse, Ragnar (1953). *Problems of Capital Formation in Underdeveloped Countries*. Oxford: Blackwell.

O'Brien, Patrick K., and Caglar Keyder (1978). *Economic Growth in Britain and France 1780–1914: Two Paths to the Twentieth Century*, London: Allen & Unwin.

O'Neil, Shannon K (2016). *This Week in Markets and Democracy: Moldova's Protests, Investors take on Graft, Corruption's Cost in Nigeria*. Council on Foreign Relations, February 5.

Odari, Edgar Jalang'o (2008). "Africa's 'Agrarian' Revolution: Land and Policy Prescriptions to Promote Impact Investment in Foreign Acquisitions." In Nhamo, Godwell, and Chekwoti (Eds.), Land Grabs in a Green African Economy: Implications for Trade, Investment and Development Policy, Pretoria, South Africa: Africa Institute of South Africa.

Okonjo-Iweala, Ngozi (2009). "Africa's Growth and Resilience in a Volatile World," *Journal of International Affairs*, *62*(2), (Spring/Summer).

Okumu, Wafula. The Role of AU/NEPAD in Preventing and Combating Corruption in Africa – A Critical Analysis, http://www.africafiles.org/printable version.asp?id=10150

Ollapally, Deepa (1993). "The South Looks North: The Third World in the New World Order." *Current History,* (April).

Oxley, J. E. (1997). "Appropriability hazards and governance in strategic alliances: A transaction cost approach." *Journal of Law, Economics & Organization*, *13*(2).

Padilla-Perez R., J. Vang, and C. Chaminade (2009). "Regional innovation systems in developing countries: integrating micro and meso-level capabilities." In Lundvall B., K. J. Joseph, C. Chaminade and J. Vang (Eds.), *Handbook of Innovation Systems and Developing Countries: Building Domestic Capabilities in a Global Setting.* Northhampton, MA: Edward Elgar.

Panda, Biswambhar (2007). "Top Down or Bottom Up? A Study of Grassroots NGOs' Approach." *Journal of Health Management*, *9*(2).

Patel, Surendra J. (1962). "Rate of Industrial Growth in the Last Century. 1860–1958." *Economic Development and Cultural Change*, (April).

Patnaik, Ila, Abhijit Sen Gupta and Ajay Shah (2010). "Determinants of Trade Misinvoicing." *Working Paper No. 75*, New Delhi: National Institute of Public Finance and Policy, (October), http://www.nipfp.org.in

Peerenboom, R. (2004). "Varieties of Rule of Law: An Introduction and Provisional Conclusions." In R. Peerenboom (Ed.). *Asian Discourses of Rule of Law: Theories and Implementations of Rule of Law in Twelve Asian Countries, France and the U.S.*

Perkins, D. W., M. Roemer Gillis, and D. R Snodgrass. (1987). *Economics of Development* (2nd ed.). New York:W. W. Norton.

Peters, Anne (2015). *Corruption and Human Rights*, Working Paper Series No. 20, Basel Institute on Governance, (September).

Peters, Dirk and Wolfgang Wagner (2010). Between Military Efficiency and Democratic Legitimacy: Mapping Parliamentary War Powers in Contemporary Democracies, 1989–2004." *Parliamentary Affairs*, *64*(1) (September). http://www.eurogendfor.org; https://en.wikipedia.org/wiki/Gendarmerie; https://en.wikipedia.org/wiki/List_of_gendarmeries; and http://www.independent.co.uk/news/world/europe/who-are-gign-elite-police-force-formed-after-1972-olympics-attack-on-israelis

Pierre J., and Peters B. G. (2000). *Governance, Politics and the State.* New York, NY: St. Martin's Press.

Pike, A., A. Rodrigues-Pose, and J. Tomaney (2007). What Kind of Local and Regional Development and for Whom? *Regional Studies*, *41*.

Pincus, J. (2009). "Vietnam: Sustaining Growth in Difficult Times." ASEAN Economic Bulletin, *26*(1).

Pingali, P. (2007). "Agricultural Growth and Economic Development: A View through the Globalization lens." *Agricultural Economics* 37(1).

Porter, M. E. (1998). "Clusters and the New Economics of Competition." *Harvard Business Review*, 76(6).

Post, Lori Ann, Amber N. W. Raile, and Eric D. Raile (2010). "Defining Political Will." *Politics & Policy*, 38(4), (August).

Prahalad, C. K., and S. L. Hart. (2004). *The Fortune at the Bottom of the Pyramid: Eradicating Poverty through Profits.* Upper Saddle River, NJ: Prentice Hall.

Prebisch, Raul (1950). "The Economic Development of Latin America and Its Principal Problems." New York: United Nations. Reprinted in 1962. *Economic Bulletin for Latin America* 7.

Putnam R., with Leonardi R. and Nenetti R.Y. (1993). *Making democracy work: Civic traditions in modern Italy.* Princeton, NJ: Princeton University Press.

Quadir, I., and G. Morse (2003). Bottom-up economics, *Harvard Business Review*, (August).

Rajagopal, B. (2003). *International Law from Below: Development, Social Movements, and Third World Resistance.* Cambridge: Cambridge University Press.

Rauchway, Eric (2008). The Great Depression & The New Deal, New York: Oxford University Press

Redding, S., and P. Schott (2003). "Distances, Skill Deepening, and Development: Will Peripheral Countries Ever Get Rich?" *Journal of Development Economics* 72(2).

Reforming the International Investment Regime: Need for African Voice, African Center of International Law Practice, September 30, 2017.

Reuer, J. J. (2001) "From Hybrids to Hierarchies: Shareholder Wealth Effects of Joint Venture Partner Buyouts." Strategic Management Journal, 22(1).

Reuer, J. J., M. Zollo, and H. Singh, (2002). Post-formation dynamics in strategic alliances. *Strategic Management Journal*, 23(2).

Revision of the AU/NEPAD AFRICAN ACTION PLAN 2010–2015: Advancing Regional and Continental Integration Together through Shared Values, Abridged Report2010-2012, NEPAD Planning and Coordinating Agency, 2011.

Robert M. Solow (1964). *Capital Theory and the Rate of Return.* Chicago: Rand McNally & Company.

Rodrigues-Pose A. and N. Gill (2006). How does trade affect regional disparities? *World Development 34.*

Rodriguez-Pose, A. (2010). Do institutions matter for regional development? *Working Paper in Economics and Social Sciences* No. 2, IMDEA Social Sciences, Madrid.

Rodríguez-Pose, Andrés and Sylvia Tijmstra (2005). *Local Economic Development as an Alternative Approach to Economic Development in Sub-Saharan Africa,* (A report for the World Bank), http://siteresources.worldbank.org/INTLED/Resources/339650–1440997 18914/AltOverview.pdf

Rodrik, D., A. Subramanian, and F. Trebbi (2004). "Institutions Rule: The Primacy of Institutions over Geography and Integration in Economic Development." *Journal of Economic Growth* 9(2).

Root, Hilton L. (1996). *Small Countries, Big Lessons: Governance and the Rise of East Asia*. New York: Oxford University Press.

Rosenau, William, Peter Chalk, Renny McPherson, Michelle Parker and Austin Long (2009) "Corporations and Counterinsurgency," Intelligence Policy Center of the RAND National Security Research Division, Santa Monica, CA: Rand Corporation.

Rosenstein-Roden, P. N. (1943). "Problems of Industrialization of Eastern and Southern Europe," *Economic Journal,* (June–September).

Rostow, W. W. (1960). *The Stages of Economic Growth: A Non-Communist Manifesto*. Cambridge: Cambridge University Press.

Roxburgh, R. (2010). *Lions on the Move: the Progress and Potential of African Economies*. Washington DC: McKinsey Global Institute.

Samaké, I. (2008). "Investment and Growth Dynamics: An Empirical Assessment Applied to Benin. Number 8–120." *International Monetary Fund*.

Sanchez-Robles, B. (1998). "Infrastructure Investment and Growth: Some Empirical Evidence." *Contemporary Economic Policy, 16*(1).

Santos, A. (2006). "The World Banks Uses of the 'Rule of Law' Promise." In David M. Trubeck and Alvaro Santos (Eds.), *The New Law and Economic Development*. Cambridge: Cambridge University Press.

Sanyal, Bishwapriya (1998). "The Myth of Development from Below." Association of Collegiate Schools of Planning, Annual Conference, Pasadena, CA. (November), web.mit.edu/sanyal/www/articles/Myth%20of20 Dev.pdf.

Schafer, S., Smith, H. J. & Linder, J. (2005). The Power of Business Models. *Business Horizons,* 48.

Schaumburg-Muller, H. (2005). "Private Sector Development in a Transition Economy: the Case of Vietnam." *Development in Practice, 15*(3)&(4).

Schultz T. W. (1961). "Investment in Human Capital." *American Economic Review,* (March).
Scott, A. J. "Regional Push: The Geography of Development and Growth in Low-and Middle-Income Countries," *Third World Quarterly*, 23.
Scott, Allen J. and Gioacchino Garofoli (2007). *Development on the Ground: Clusters, Networks and Regions in Emerging Economies*. New York: Routledge.
Seidman, R. and A. Seidman (1999). "Using Reason and Experience to Draft Country-Specific Laws." In A. Seidman, R. B. Seidman and T. Walde (Eds.), *Making Development Work: Legislative Reform of Institutional Transformation and Good Governance*. London: Kluwer Law International.
Sen, Amartya (1999). *Development as Freedom*. New York: Random House.
Serfaty, Simon (2012). *A World Recast: An American Moment in a Post-Western Order*. New York: Rowman & Littlefield.
Seth, Tushar. *Reasons for Low Capital Formation in Under-Developed Countries*, http://www.economicsdiscussion.net/articles/reasons-for-low-capital-formation-in-under-developed-countries/1537
Shuaib, I. M., and Evelyn Ndidi Dania (2015). "Capital Formation: Impact on the Economic Development of Nigeria 1960–2013." *European Journal of Business, Economics and Accountancy*, 3(3).
Singer, Hans (1950). "Gains and Losses from Trade and Investment in Under Developed Countries." *American Economic Review* (May).
Singer, Hans W., and Javed A. Ansari (1988). *Rich and Poor Countries: Consequences of International Disorder* (4th ed.). Boston: Unwin and Hyman.
Singh, Chaitram (1989). *Multinationals, The State, and The Management of Economic Nationalism: The Case of Trinidad*. New York: Praeger.
Sintim-Aboagye, Hermann (2012). "IMF and World Bank Economic Programs on Inflation: Lessons for NEPAD." In Benjamin F. Bobo, and Sintim-Aboagye (Eds.), *Neo-Liberalism, Interventionism and the Developmental State: Implementing the New Partnership for Africa's Development*. Trenton New Jersey: Africa World Press.
Smith, Adam (1937). *An Inquiry into the Nature and Causes of the Wealth of Nations* (1st ed.). 1776, New York: Modern Library.
Smith, Adam (1978). *The Wealth of Nations*. Franklin Center, Pennsylvania: The Franklin Library, A Limited Edition.

Smith, Adrian, M. Fressoli, H. Thomas, and D. Abrol (2012). "Supporting Grassroots Innovation: Facts and Figures." *Sci Dev Net.* http://www.scidev.net/global/icts/feature/supporting-grassroots-innovation-facts-and-figures-1.html, February 5.

Solow R. (1956). A contribution to the theory of economic growth. *The Quarterly Journal of Economics 70.*

Somerville, Peter (2016). *Understanding Community: Politics, Policy and Practice.* University of Bristol: The Policy Press.

Stern, David I. (1997). "The Capital Theory Approach to Sustainability: A Critical Appraisal." *Journal of Economic Issues, XXXI*(1), (March).

Streeten, P. P. (1967). "The Frontiers of Development Studies: Some Issues of Development Policy." *Journal of Development Studies,* (October).

Sundaram, Jomo Kwame (2007). "What did we really learn from the 1997–98 Asian debacle?" In Bhumika Muchhala (Ed.), *Ten Years After: Revisiting the Asian Financial Crisis*, Washington, DC: Woodrow Wilson International Center for Scholars.

Tammen, Ronald L., et al. (2000). *Power Transitions: Strategies for the 21st Century.* New York, NY: Seven Bridges Press.

Tang, Min, and Dwayne Woods (2008). "The Exogenous Effect of Geography on Economic Development: The Case of Sub-Saharan Africa." *African & Asian Studies, 7*(2/3).

The Cost of U.S. Overseas Military Bases (2012). LessWaiting.com, April 23. http://www.globalsecurity.org/military/facility/images/diego-garcia-ims7.jpg.

Tiffen, M. (2003). "Transition in Sub-Saharan Africa: Agriculture, Urbanization and Income Growth." *World Development 31*(8).

Tipton, F. B. (2009). "Southeast Asian Capitalism: History, Institutions, States and Firms." *Asia Pacific Journal of Management, 26.*

Tran-Nam, B. (2005). "Globalization, Trade, Liberalization and Economic Growth: The Case of Vietnam." In M. Chatterji, and P. Gangopadhyay (Eds.), *Economic Globalization in Asia.* New York, NY: Routledge.

Transfer Pricing Updates (2014). "Across Africa: EY Africa Tax Conference, September," EY: Ernst & Young Global Limited, ey.com,

Transparency International (1998). *Corruption-A Violation of Human Rights?* Transparency International working paper. International Council on Human Rights.

Transparency International (2006). *Towards a Fairer World: Why is Corruption Still Blocking the Way? Goals, Themes, and Outcomes.* Transparency International. www.12iacc.org/archivos/12IACC-

Transparency International (2007). *Global Corruption Report 2007: Corruption in Judicial Systems*. Cambridge: Cambridge University Press.

Transparency International (2009). *Corruption and Human Rights. Making the Connection*, International Council on Human Rights Policy, Versoix, Switzerland.

Transparency International (2015). Corruption Perceptions Index 2015, www.transparency.org/cpi2015

Transparency International (the global coalition against corruption), http://www.trans parency.org/whoweare/organisation/faqsoncorruption

Trubeck, D. M. (2006). "The 'Rule of Law' in Development Assistance." In David M. Trubeck and Alvaro Santos (Eds.), *The New Law and Economic Development: A Critical Appraisal*. Cambridge: Cambridge University Press.

Tsheola, Johannes (2010). "Global 'openness' and trade regionalism of the New Partnership for Africa's Development," *South African Geographical Journal*, 92(1).

Udombana, Nsongurua J. (2003). "Fighting Corruption Seriously? Africa's Anti-corruption Convention." *Singapore Journal of International & Comparative Law*, 7.

ul Haq, Mahbub (1995). *Reflections on Human Development*. Oxford: Oxford University Press.

UNCTAD (1978). *Dominant Positions of Market Power of Transnational Corporations: Use of the Transfer Pricing Mechanism*. New York: United Nations.

UNCTAD (2009). *The Role of International Agreements in Attracting Foreign Direct Investment to Developing Countries*. New York: United Nations.

UNCTAD (2014). "Catalysing Investment for Transformation Growth in Africa." Economic Development in Africa Report 2014 UNCTAD/ALDC/AFRICA/2014/5114, United Nations Conference on Trade and Development. New York: United Nations.

UNCTAD (2014). *Economic Development in Africa Report 2014*. UNCTAD/ALDC/AFRICA/2014, New York: United Nations.

Uneze, E. (2013). The Relation between Capital Formation and Economic Growth: Evidence from Sub-Saharan African Countries. *Journal of Economic Policy Reform* 16(3).

United Nations Economic Commission for Africa, http://www.uneca.org/

Walleri, Don (1978)."The Political Economy Literature on North-South Relations," *International Studies Quarterly*, 22(4), 589.

Wenger, Andreas, and Daniel Mockli (2003). *Conflict prevention: The Untapped Potential of the Business Sector*. Boulder, CO: Lynne Rienner Publishers.

Wholstetter, A. J. (1951). *Economic and Strategic Considerations in Air Base Location: A Preliminary Review*. Santa Monica, CA: RAND Corporation, D-1114, 29 December.

Williams, Jeremy (2007). *Make Wealth History*, https://makewealthhistory.org/2007/11/05/whyare-some-countries-poor/

Williams, Walter E. (2014). "Africa: A Tragic Continent. But Why?" *The New American*, (October 29).

Willis, Katie (2005). *Theories and Practices of Development*. New York: Routledge.

Wohlstetter, Albert and Harry Rowen (1952). *Campaign Time Pattern, Sortie Rate, and Base Location*. Santa Monica, CA: RAND Corporation, January.

Wolf, Charles (2013). "A Truly Great Leap Forward" in *The Rand Blog*, http://www.rand.org/blog/2013/05/a-truly-great-leap-forward.html. Santa Monica, CA: The Rand Corporation.

Wong, C. (2002). Developing Indicators to Inform Local Economic Development in England. *Urban Studies 39*.

World Bank (1989). *Sub-Saharan Africa: From Crisis to Sustainable Development*. World Bank.

World Bank (1998), *Assessing Aid: What Works, What Doesn't and Why*. Oxford: Oxford University Press.

World Bank, *Fact Sheet: Infrastructure in Sub-Saharan Africa*, http://go.worldbank.org/SWDECPM5S0

World Factbook (2015). Washington, DC: Central Intelligence Agency (USA).

Wrigley, E. A. (1987). *People, Cities, and Wealth: The Transformation of Traditional Society*, Oxford: Basil Blackwell.

Yang, Yongzheng, and Sanjeev Gupta (2007). *Regional Trade Arrangements in Africa: Past Performance and the Way Forward*. African Development Bank: Blackwell Publishing Ltd.

Yeaple, Stephen Ross (2003). "The Complex Integration Strategies of Multinationals and Cross Country Dependencies in the Structure of Foreign Direct Investment." *Journal of International Economics*, 60(2), (August).

Zaaier, M., and L. M. Sara (1993). "Local Economic Development as an Instrument for Urban Poverty Alleviation: A Case from Lima, Peru." *Third World Planning Review 15*.

Zhang X., and Zhang K. (2003). "How does globalization affect regional inequality within a developing country? Evidence from China." *Journal of Development Studies, 39*.15 (February).

Index

A
AU 112, 153, 159, 189
Achilles' heel 3, 45
Adam Smith 4, 171
Africa 9, 23, 117, 185
Africa MNC Strategic Alliance Decision-Making Protocol 8, 197
Africa Trade Fund 175
Africa's predicament 25, 142
Africa-MNC alliance 4, 121, 166, 185
Africa-MNC Strategic Alliance 8, 117, 137, 182
African branding 136
African Central Bank 46
African continent 1, 80, 138, 185
African Continental Security Apparatus 185
African countries 1, 86, 131, 196
African Development Bank 44, 106, 154, 176
African Regional Integration Index 154
African Security Force 139, 176
African Union 1, 47, 112, 197
African Union's Peace and Security Council 8, 113
Agriculture sector 1
Agriculture 1, 29, 78, 157
Albert O. Hirschman 27

Algeria 39, 86, 102, 223
Alliance Operations 118, 174
Amartya Sen 47
Analytical policy framework 23
Andre Gunder Frank 27, 145, 169, 180
Anglophone 136
Angola 39, 86, 98, 223
Antonio Costa 130
Ascani 59, 60, 61, 74
Asia 18, 23, 65, 159
Asian debacle 20, 21
Assisted capitalism 49
Assisted commerce 49
AU/NEPAD African Action Plan 137, 141, 153, 159
Auditing 132

B
Bangladesh 65, 115, 164
Banks 52, 62, 64, 66
Basdeo 112, 114, 115
Benin 40, 86, 101, 223
Benjamin F. Bobo 14, 53, 107, 180
Bernard Madoff 13
Bertil Ohlin 151
Bhumika Muchhala 21
Bidder 12
Big push thesis 27
Bilateral investment treaties 111

Bodies of government 17, 18, 20, 138
Botswana 40, 86, 100, 223
Bottom Up 57, 69, 121, 139
Brazil 43
Buehn 168
Bundle of rights theory 76
Burkina Faso 39, 77, 86, 223
Burundi 39, 87, 223

C

Cadre of advisory specialists 2
Cameroon 39, 77, 87, 223
Canada 25, 39, 104, 156
Capacity development 51
Capacity ID 5, 85, 140, 177
Capacity ID Attributes 84, 140, 177
Capacity Innovation 23, 57, 137, 165
Capacity innovation endgame 49, 186
Capacity innovation scheme 1, 25, 134, 140
Cape Verde 40, 87, 100, 223
Capital Diffusion 54, 112, 139, 182
Capital formation 53, 86, 102, 174
Capital theory 9, 11, 122, 196
Capital-intensive 32, 145
Capitalism and freedom 15, 22, 59, 122
Capitalism-led development 17
Capitalist institution 10, 14, 16
Capitalist model 16, 122
Catch-22 45, 46
Celestous Juma 1
Celso Furtado 27
Central African Republic 87, 223

Chad 39, 87, 223
Chairman Mao 23
Chaitram Singh 178, 179
Change agent 47, 73, 138, 172
Characteristics of the Multinational Corporation 143
China 23, 66, 114, 169
Chinaphone 136
CIA World Factbook 99
Citizen Participation 84, 108, 109, 116
Classical theory of economic growth 27
Collective self-reliance 22, 37, 44, 138
Communication 36, 43, 139, 216
Comoros 40, 87, 103, 223
Comparative advantage 33, 37, 150, 186
Competencies 1, 7, 61, 144
Computer-based networking 45
Congo, Democratic Republic 39, 88
Congo, Republic of 88
Conover 57, 59
Constitution 18, 86, 107, 138
Continuity/discontinuity 27
Continuity/periodization 27
Conventional wisdom 2, 123, 185
Convergence 69, 72, 139, 163
Coordination 71, 73, 200, 219
Core capacity attributes 86, 107, 115, 116
Core challenges 36, 66, 138, 186
Corrupt officials 15, 16
Corruption dilemma 124
Cost-effective 2, 62, 63
Cote d'Ivoire 39, 88

Country capacity ID 83, 116, 140, 177
Created comparative advantage 33, 138, 150, 182
Creating regional markets 1
Crescenci 70, 71, 72
Crimes against humanity 127, 134, 197, 204
Critical mass 36, 37, 138, 164
Cross-fertilization 72, 74, 148, 197
Cross-fertilizing 71, 118, 140, 148
Cross-fertilizing integrative framework 117, 140
Currency 36, 45, 104, 175
Curry 178, 179
Curry-Rothchild Model 179

D
David I. Stern 9
David Landes 55
David N. Balaam 36, 37, 145, 154
David Ricardo 27
Debate 9, 75, 100, 150
Democratic Republic 88, 92, 223, 225
Dependency approach 27
Dependency theory 27, 142
Destructive rent-seeking 76
Development failure 26, 55
Development tools 1, 7
Developmental state 14, 53, 106, 140
Distrubutional capacity 2
Djibouti 40, 89, 223
Doi moi 73
Dominate society 28, 30
Drive to Maturity 27, 28, 31, 32

E
E. Wayne Nafziger 27, 174
East Timor Truth Commission 133
Easterly 67, 68, 76, 77
Economic Commission for Africa 105, 106, 154, 155
Economic Community of West African States 44
Economic complementarity 36, 42, 139, 154
Economic drivers 2, 5, 7, 196
Economic innovation 42, 51, 54
Economic integration 36, 44, 139, 159
Economic integration agreements 111
Economic sectors 2, 32
Economic structural arrangement 2, 196
ECOWAS 44
Edgar Jalang'o Odari 79
Efficacy 1, 62, 107, 146
Efficiency-equity tradeoff 80
Efficient 36, 61, 140, 191
Efficient market mechanism 22, 30, 51, 63
Efficient outcome 2, 49, 115, 163
Egypt 38, 39, 89, 224
Eichler 168
El Salvador 114
Endgame perspective 49, 50
Enforcing efficiency 2
Enron Corporation 13, 15
Entrepreneurship 31, 63, 131, 174
Epicenter of economic failure 26
Equatorial Guinea 40, 89, 100, 224

256   *Index*

Eric Rauchway 21
Eritrea 40, 89, 224
Ethiopia 39, 89, 224
Europe 25, 36, 43, 159
Executive 20, 110, 132, 220

F

FDI 3, 24, 84, 109
FDI Fund 119, 173, 174, 192
FDI Protection 139, 176
FDI Security 112, 186, 187, 192
Federal Reserve 20, 21
Fife 73, 158
Fifty-five countries 35
Firestone in Liberia 176
First World 5, 10, 49, 141
Fitzroy 148
Five-stage schema 35
Foreign Direct Investment 5, 79, 109, 138
Foreign Exchange 46, 83, 168, 175
Foreign exchange reserve fund 46, 47, 174, 175
Framework for African Capacity Innovation 23, 49, 57, 137
France 25, 39, 114
Francophone 136
Frank F. K. Byamugisha 77, 78
Free enterprise 17, 20, 51, 62
Free market 12, 67, 140, 151
Free market system 16, 17
Free-rider problem 46, 58, 175
Friedman 14, 15, 63

G

G8 countries 25, 31, 41, 98
Gabon 40, 90, 98, 224
Gambia, The 90
GARCH model 147
Garofoli 164, 165, 166
GDP 38, 101, 150, 180
GDP per capita 38, 74, 101, 166
Gendarmerie 86, 112
General Hoyt Vandenberg 190, 191
Gerald M. Meier 23, 28, 31, 84
Germany 25, 39, 114
Ghana 39, 78, 90, 224
GIST 170, 171, 178, 179
Give-and-take 16, 17, 51
Global financial integrity 10, 132, 167, 169
Global Interdependency Sensitivity Thesis 170, 178, 179
Good governance 18, 36, 126, 206
Governance 36, 47, 126, 203
Government apparatus 17, 18, 20
Government corruption 11, 14, 16, 21
Government institutions 5, 30, 140, 217
Governmental decree 17, 18
Granger Causality 147
Grassroots 5, 69, 138, 197
Great depression 20, 21
Guinea 39, 90, 101, 224
Guinea-Bissau 40, 91
Gunnar Myrdal 84, 85

H

Hard currency 37, 45, 121, 174
Hermann Sintim-Aboagye 14, 53, 106, 146
High mass consumption 27, 28, 32

Highest and best use 16, 31, 76, 173
Hilton L. Root 18, 47, 107
Historical materialism 27
Holistic approach 6, 56, 179, 196
Hosman 73, 158
How Africa can prosper 36, 43, 45, 154
Human development index 73, 88, 101, 102
Human development paradigm 73, 125
Human rights violations 126, 127, 128

I

Iammarino 59, 60
ICT 158
Illiteracy 26
IMF 7, 138, 146, 181
Import substitution 36, 44, 121
Import/export 46, 175
India 26, 65, 127, 164
Indigenous stakeholders 79, 80
Industrialized Nations 26
Infrastructure 43, 84, 105, 153
Infrastructure development 7, 83, 138, 159
Institutional capacity 7, 48, 124, 216
Institutional component 138, 141
Institutional mapping 18, 48, 84, 141
Integrative 69, 71, 117, 140
Integrative approach 69, 71
Interdicting Corruption 134
International currency 36, 45
International investment agreements 111

Intervention 37, 53, 140, 175
Intra-Africa communications systems 36
Intra-African currency 46, 174
Intra-African trade 42, 45
Intra-continental communication 53, 139, 159
Invisible hand 4, 49, 124, 171
Italy 25, 38, 39, 114

J

Jamee Moudud 14, 51, 106
James A. Brander 14, 151
Japan 25, 38, 114, 191
Jeremy Williams 195
Jomo Kwame Sundaram 21
Judicial 13, 19, 20

K

Kar and LeBlanc 10
Karl Botchway 14, 51, 106
Karl Marx 27
Kenya 68, 78, 91, 101
Kenyan 68, 76
Kickbacks 15, 21
Kigali Protocol 158
Kipaya Kapiga 76
Klien 68
Knowledge-intensive 32, 35, 41, 49
Kohler 113, 114
Kraft 148

L

Labor-intensive 31, 35, 41, 156
Ladder of comparative advantage 31, 70, 138, 186
Laissez-faire 51
Land tenure 75, 76, 80

Latin America 23, 43, 65
Law 15, 18, 36, 77
Leading Issues in Economic Development 23, 29, 32, 151
Leading sectors 27
LeBlanc 10, 167
Legislative 13, 20
Lesotho 40, 91, 100, 224
Liang 114
Liberia 40, 91, 176, 224
Libya 40, 91, 102, 224
Linking process 141, 178, 179, 182
Locked In and Locked Out 70, 85, 180
Locking effect 85, 180
Lu 114

M

Madagascar 39, 77, 91, 224
Mahbub ul Haq 11, 73, 125
Malawi 39, 92, 224
Malfeasance 9
Mali 39, 92, 201, 224
Market component 138, 142
Market deficiencies 1, 6, 7
Market selection 2, 5, 13, 196
Maturity 27, 28, 31, 32
Mauritania 40, 99, 103, 224
Mauritius 92, 101, 103, 106
Mediterranean 29
Member states 125, 189, 216, 223
Michael Veseth 36, 145, 154
Middle East 29
Miethe 114
Milton Friedman 14, 15
Mini-states 36, 37, 38, 164
Misfeasance 9

Missing middle 73
MNC 71, 73, 121, 139
MNC-NEPAD Committee 174
Mobile Cellular Market 43
Modernity 30, 56, 138, 197
Modernization 4, 30, 125, 178
Montesh 112, 114, 115
Morocco 39, 43, 92, 103
Mozambique 39, 93, 224
Multinational corporation 5, 172, 186, 198
Mutually exclusive 112

N

Namibia 40, 93, 100, 104
NATO Alliance 191
Natural comparative advantage 33, 41, 154, 157
Natural resources 31, 135, 152, 158
Nee 58
Neo-liberal 21, 49, 51, 140
Neo-Marxist thesis 27
NEPAD 1, 25, 140, 164
NEPAD Bank 139, 175, 182
NEPAD-MNC-NGO Convergence 163
New Deal 21
New Partnership for Africa's Development 1, 3, 110, 126
NGO 5, 64, 65, 118
NGOs 62, 65, 75, 121
Niger 39, 93, 101, 225
Nigeria 38, 41, 93, 177
Njoroge 129, 130, 133
Nobel Laureate 47
Non-governmental organizations 65, 120, 147, 221

Non-market selection 2
Nourse 157

O

OECD 78, 109, 161
One-size-fits-all 61, 146
OPEC 41
Opportunistic balancing 5, 121, 134, 178
Opportunistic behavior 13, 76, 120, 149
Organization for Economic Cooperation and Development 78, 145
Organizational tie-in 140, 178, 179, 183

P

Pak Hung Mo 9, 11
Panel of advisors 2
Path to capacity innovation 5, 52, 53, 137
Patnaik 168
Paul A. Baran 27
Paul N. Rosenstein-Rodan 27, 165
Paul R. Krugman 1, 14, 150, 151
Peace and Security Council 115, 179, 188, 201
Peace and Security Council Protocol 186
Persian Gulf 191
Philippines 80, 114
Placer Dome in Papua New Guinea 176, 177
Population 25, 40, 98, 110
Post-subprime 34
Power principle 18, 20
Power-dominant attributes 83, 84

Preconditions for take-off 27, 28, 29, 33
Preferential trade and investment agreements 111
Preferential Trade Area 44
Pre-Newtonian 29
Price-maker 7, 36, 139, 156
Price-maker signaling 7
Price-taker 36, 37, 41
PricewaterhouseCoopers 131, 132
Private markets 5, 9, 14, 196
Process 3, 28, 122, 141
Production functions 28, 29, 30
Property rights 58, 75, 81, 124
Protect foreign direct investment 7, 83, 114, 185
Protocol Relating to the Establishment of the Peace and Security Council of the African Union 113, 127, 185, 199

Q

Quinlivan 143, 145

R

R&D 32, 35, 41, 49
Rand 11, 66, 113, 176
Rate of investment 30, 103
Regional Integration 45, 53, 138, 152
Rent seeking 76, 171
Resource allocation 11, 51, 140, 196
Resource-intensive 31, 33, 41, 49
Resources of commercial importance 86, 97, 98, 115

Retribution 13
Rich country, Poor country 47, 74, 143, 172
Robert Curry 178
Robert S. Browne 36, 37, 154
Robert Solow 11
Rodriguez-Pose 69, 70, 72, 74
Role of the State 17, 21, 58, 146
Role players 12
Ronald Brand 111
Rosenau 176, 177
Rothchild 178, 179
Royal Dutch Shell in Nigeria 176, 177
Rule of law 15, 20, 53, 84
Russia 25, 38, 39
Rwanda 39, 93, 225

S

Sanyal 64, 65, 118, 164
Sao Tome and Principe 94, 100, 225
School of thought 51
Scope 10, 43, 115, 169
Scott 164, 165, 166
Sectoral selection 2
Security apparatus 8, 111, 185, 186
Security Hubs 189, 190
Senegal 39, 94, 146, 225
Serfaty 165
Seychelles 40, 94, 102, 225
Sierra Leone 40, 94, 101, 225
Small Countries, Big Lessons: Governance and the Rise of East Asia 18, 47, 107
Socio-economic construct 51
Solyndra LLC 13
Somalia 40, 94, 101, 225

South Africa 39, 74, 99, 106
South African Defense Act 112
South African National Defense Force 112, 115
South African Police Service 112
South Korea 191
South Sudan 39, 95, 101, 133
Special fund 174, 192, 204, 221
Special interest 9, 18, 102, 219
Stage discerning/periodization issues 27
Stages of economic growth 27, 28, 70, 148
Stakeholder givebacks 138, 148, 149, 174
Stakeholders 58, 79, 149, 162
Stiglitz 50
Strategic alliance 8, 81, 117, 137
Strategic banking scheme 7
Strategic Readiness Response 189, 190
Strategy of unbalance 27
Subprime fiasco 20
Sudan 39, 95, 101, 225
Supply-side constraints 1, 7, 137, 156
Supporting institutions 15, 16, 17
Sustained economic transformation 1, 2, 8
Swaziland 40, 95, 100, 225
Synergistically 74, 164, 197
System of institutions 15, 16, 17

T

Taiwan 191
Take-off 12, 28, 72, 137
Tanzania 39, 96, 225
Targeting 1, 52, 53, 150
TARP 21

Thabo Mbeki 47
The New Partnership for Africa's Development 14, 53, 126, 140
The system 64, 76, 157, 180
The Wealth of Nations 4, 171
Third World 23, 45, 145, 195
Togo 40, 96, 125, 225
Top Down 57, 69, 72, 138
Top down-bottom up with a twist 71, 75
Trade misinvoicing 139, 166, 168, 169
Trade theory 2, 122, 142, 150
Traditional society 27, 28, 29, 35
Transfer pricing 139, 166, 167, 170
Transparency International's Corruption Perceptions Index 130, 133
Triple alliance 118
Troubled Asset Relief Program 21
Trump Card 134, 135
Tunisia 39, 96, 102, 225
Twist 71, 141, 174, 196

U
UK 25, 40, 114
US 23, 25, 114, 191
Uganda 39, 96, 146, 225
UNCTAD 105, 111, 167
United Nations Economic Commission for Africa 106, 154
Unlocking the lock 140, 180
UNODC 130

V
Vietnam 58, 59, 73

W
Walter W. Rostow 27
Weak-deficient attributes 84
Western Sahara 40, 96, 105, 115
William Easterly 67, 76
Winners and losers 1, 13, 59, 124
Workforce skills 1, 6, 7
World Bank 36, 69, 138, 146
World Commission on Environment and Development 9
Wrongdoers 13, 15, 16

Y
Yang and Gupta 42, 106
Yield 31, 110, 143, 169

Z
Zambia 39, 96, 199, 225
Zimbabwe 97, 100, 132, 225

About the Author

Benjamin F. Bobo is a professor of finance, College of Business Administration, Loyola Marymount University where he teaches finance at the undergraduate level and international economic strategy and trade policy in the graduate program. He has held faculty posts at the University of California, Los Angeles and the University of California, Riverside and has served as Assistant Secretary for Policy Development and Research (Acting) at the U.S. Department of Housing and Urban Development in Washington, D.C. He has headed U.S. delegations to the former Soviet Union, Switzerland, Greece, Mexico and Canada.

Dr. Bobo has implemented U.S. policy on international housing technology and has served as Co-Chair of Joint Steering Committees of U.S. bilateral agreements. He has served as consultant to the U.S. Department of Commerce and to private firms, and engaged business development activities in Africa, Europe, and the Caribbean. He has published extensively on Third World development; his research focuses on life-choice constraints of the economically disadvantaged.